Flight in Jersey
The Story of Jersey Airport

The airport as it was in March, 1987, the year it celebrated its 50th anniversary

Prediction

ON Saturday, October 20, 1917, Miss Gertrude Bacon, the first woman to travel by airship, gave a talk to the St Helier's Church Literary Society entitled 'Flying Machines In Peace And War'. She made two predictions.

First, that after the Great War, civilians would travel from London to Paris by air.

Second, that an airport would eventually be built in Jersey. At a time when motor cars were viewed as novelties, seven years before the first electricity cable was laid in St Helier and at a time when most islanders rarely moved from one parish to another, let alone travelled to London or Paris, Miss Bacon's prophecies seemed wildly audacious.

However, beneath her insistence that regular air travel was inevitable and that an airport would be built in Jersey (for dirigible balloons, she believed, not aeroplanes), there was a great deal of truth in what she said. On Wednesday, March 10, 1937, at 2.45 pm, prompt, the Island's civic head, the Bailiff, Mr A M Coutanche, made a short speech before his wife officially opened the new Airport at St Peter.

By 1997 over a million and a half passengers were passing through its glass doors each year. Had Miss Bacon ever envisaged so many people arriving in the Island, by air?

February 15, 1997: Less than a month to go before the new departures hall will open

'Flight in Jersey' (the Story of Jersey Airport) was first published as 'Jersey Airport: The First 50 Years' in 1987. Now, as then, the publishers would like to thank all those people involved with the preparation of this book, including the many photographers whose pictures were loaned by private collectors. The book would have been impossible to compile without the help, a decade ago, of the following: Jersey Airport, Airport Administration, J Beuzeval, Senator B Binnington, P Cotillard, D Ford, W M Ginns, Hawker Siddeley Company, J Herbert MBE, K Jackson, Jersey Evening Post, Société Jersiaise, E Knight, M R Lanyon, I Law, K J Le Cocq, D Luke, R Mayne, L Daligault, J Hadley, Postal Administration, RAFA (Jersey Branch), RAF Museum (photographic section) Hendon, S Senett, L Sinel, R Wagstaffe, P Law, L Le Moignan, Jersey Museum, Musée de l'Air, Paris.

Cover design by Mike Warstat
Written and updated by Chris Lake
New colour photographs by Gary Grimshaw
Printed by Grange Press, Southwick, West Sussex

Contents

AVIATION and the horse and carriage era came face to face in Jersey on August 31, 1912, when Jean Benoist landed his Sanchez-Besa biplane in St Aubin's Bay to lead the first leg of the St Malo—Jersey Hydro-aeroplane contest. Benoist eventually had to settle for third place having been delayed returning to St Malo by a large crowd which had never before seen an aircraft. The winner was another Frenchman, Weymann, with Molla second and Renaux fourth. Labouret was overall winner of three days of racing. It was to be another 25 years before Jersey had an airport on dry land.

'What a marvellous adventure Jersey has had with aviation! Less than ten years after the Wright brothers' first flight in America, aircraft landed on Jersey's beaches. Twenty years later the States of Jersey were in the forefront of aviation-minded governments by building an airport. Approaching the Millennium Jersey Airport is 'still going strong' and playing a vital part in the economic and social development of the Island. With nearly one million arriving and one million departing passengers annually, the airport is the premier gateway to the Island for business travellers and tourists alike.

' The marvellous new facilities provided in the John Le Fondré Departures Hall provide passengers a high degree of comfort, plenty of space and, with a dozen shops, opportunities to shop in a wide range of retail facilities. During the winter Jersey is served by 30 airport connections within the United Kingdom and the rest of Europe and, in the summer, travellers from 80 airports can fly directly to Jersey.

'Truly, Jersey Airport is the Gateway to the Island.'

— Michael Lanyon, Jersey Airport Director, March, 1997

The Early Days

NEARLY 200 years ago, Jersey witnessed a man crossing the Island by air when, in June, 1790, Mr Granger, a local schoolmaster, took off in a balloon from the courtyard of the Hospital. A week later he enjoyed another flight and, after taking off from La Mont de la Ville (Town Hill), spent 45 minutes in the air before dropping gently to the sea, where he was safely picked up.

Air travel had arrived in Jersey, but another 80 years were to elapse before this novel way of travelling was used for any practical purposes, when a letter arrived from Paris, carried by a man in a hot air balloon from the centre of the capital to its outskirts, and thence onwards by surface transport.

The flight caused some temporary excitement but another, heavier form of transportation was of more local interest in 1870. The railway had arrived. Initially it only stretched between St Helier and St Aubin but by 1900 you could take a train from Corbière, in the west, to Gorey Harbour, in the east (change at the Weighbridge and Snow Hill).

The last piece of railway track, between La Moye and Corbière, was laid by Jersey Railways and Tramways Ltd in 1899, the year that the first motor car arrived in the Island. It was a Benz (top speed 9.3 mph) which arrived from the mainland in July, 1899.

Four years later, in December, 1903, in America, the Wright brothers took their Wright Flyer (1) into the air at speeds approaching 30 mph and, on Monday, August 26, 1912, the first aeroplane landed in Jersey. It was a Sanchez-Besa biplane, one of four seaplanes taking part in a St Malo-Jersey race.

The event, organised jointly by the Automobile club of France and the Jersey Motor Association, drew thousands of Islanders to the sands at St Aubin to watch. Most of them had never seen a flying machine before, so to watch a hydro aeroplane (the prefix 'hydro' was eventually dropped) land and then take off again was a novel experience and one for which the Island authorities were ill-prepared.

As Benoist, the pilot of the Sanchez-Besa, flew across the sea and drew the plane up on to the sands near West Park, a mass of onlookers surged forward to poke and prod at this new-fangled invention.

Meanwhile, the refuelling vessel, moored out to sea, had to hurry back to shore before fuel could be transported to the French pilot's plane. Benoist was not a happy man by this time. He was appaled by the behaviour of the Jersey people, as they pushed against the frame or scratched their names in the wings. It wasn't their fault — they hadn't seen an aeroplane before.

In a desperate attempt to keep the crowds away, Benoist hurled nuts and bolts towards them as one of his two passengers poured petrol over those spectators closest to the cockpit. It took a long time to refuel the plane and to clear a path so that it could scud across the sand, back into the water and up into the sky, to St Malo. Benoist's delay meant that he lost the race.

An American, Charles Weymann, who had landed at Beaumont (along the beach and away from the crowd) stayed in the Island for a little over half an hour before making his way back to St Malo. He arrived at the French seaport at   11.06 am, having averaged a speed of just under 60 mph for the 90-mile course.

The age of the plane had come to Jersey, a fact which was soon recorded in the States (the Island's government) records, for on May 24, 1913, an amendment to the Army Act was produced in the Royal Court. From henceforth all types of aircraft could be requisitioned in times of need.

It was an interesting amendment to make — particularly as no one (at the time) had any aircraft to requisition, anyway. Two years after Weymann won the St Malo to Jersey race, the islands were

Benoist was worried that over-eager crowds would damage his machine

caught up in the Great War.

In July, 1914, the Royal Jersey Militia were mobilised and in February, 1917, compulsory military service was introduced. In all, 6,292 Jerseymen (and women) fought with the allies and, of these, 179 were involved with the Royal Flying Corps, the Royal Naval Air Service or the Royal Air Force.

Of those islanders who were involved with the military air services, eight died — a small number when compared with the other 854 people who never returned to their island homes. However, the number of islanders who had been involved, in some way or another, in this increasingly important form of transport, gives some indication of how, slowly but surely, aircraft were making their mark on the world. No one, least of all the French, underestimated the role that pilots might have in technological age. It had been the French and not the English who had invested a great deal of money in aircraft development in the run-up to the Great War (before 1914 they had three times as many dirigibles and hydroplanes as Britain) and it was the French government which decided to establish a seaplane station off Castle Cornet, in Guernsey's St Peter Port, after German submarines had been spotted in Channel waters in 1916.

Controlled by the Aviation Maritime Française, the Channel Islands' first 'airport' employed 110 men who serviced or flew a variety of amphibious bi- and tri-planes. (Throughout this book the term 'amphibian' is used, as in contemporary records, to refer to planes which could land on the sea. only some of them, like the Saro Cloud, had retractable wheels and could also alight on land, making them true amphibians.)

Meanwhile Jersey slumbered, although islanders may have been aware of the French presence in the air, particularly on days like April 23, 1918, when French reconnaissance planes spotted and then attacked enemy U-Boats which were hiding a little way out to sea.

Such attacks were comparatively one-sided and there was no need for the French to land in Jersey, and between 1914 and 1920, two years after the Great War had ended, only a few intrepid aviators called on the Island.

They included the unexpected arrival of Lieutenant Sidney Burgess, who made a

The 1920s: An early Imperial Airways seaplane leaves St Helier Harbour ready for take-off

1926: An Imperial Airways pilot in flying gear

forced landing at Grève de Lecq when he discovered that one of the horns on his elevator control was bent, and whose three nights in Jersey were filled with anxiety as his damaged plane became even more damaged as it was towed to Cherbourg. Lieutenant Burgess arrived on May 29, 1918.

Two months afterwards Flight-Lieutenant Stanley Mossop landed his Wight seaplane in the Harbour so that he could spend an afternoon with his family before flying back to Cherbourg.

Some 14 months after that a Mr H Storey flew a consignment of newspapers to the Island because of a strike on the mainland by the National Union of Railwaymen. The 'Seaplane Special', a 130 hp Avro biplane loaded with copies of Lloyds Weekly News, made a successful landing on October 5, 1919, and the newspapers were eagerly bought by the islanders although this successful newspaper story wasn't repeated in Guernsey, where a sister plane was due to land the same day.

The pilot of the other plane found his way from Southampton to the Channel Islands without any trouble but, as he approached Alderney, engine trouble

February, 1997: Sixty-four years after the airport was the harbour, a view from the air showing it now

A Supermarine Southampton which brought Air Chief Marshal Ellington to Jersey in 1933

The ill-fated Cloud of Iona in St Helier Harbour

forced him to come down in the sea. Plane, newspapers (but luckily not the pilot) were all lost. Mercifully, tragedy had been averted, although the story of one of the other pilots mentioned earlier in the chapter doesn't have such a happy ending. For on August 10, 1918, Stanley Mossop was only two days away from his death.

It was an unofficial visit he made to his parents home at Commercial Buildings, St Helier, that day and after taxi-ing in and leaving his plane moored near the Harbour he spent some time in St Helier before taking off again, for France. Two days later, on August 12, he (and his navigator) crashed to their deaths in the harbour of Port-en-Bessin, Calvados, while trying to land with a damaged tailplane. Mossop was just 19.

For the first 20 years of the 20th century, therefore, the impact of these new flying machines on Island life was minimal and although islanders were intrigued by aeroplanes, they often viewed them with deep suspicion.

There was an added problem that any entrepreneur who contemplated setting himself up in the aviation business had to consider. To whom would his aircraft appeal? And if they appealed, could they make him money?

The Cloud of Iona on the beach at West Park

A Most Peculiar Sensation

'I'M probably the only islander alive who remembers what it was like to fly in the Cloud of Iona. I've got a picture of her somewhere. I flew on her twice, once on July 19, 1936, to Alderney, and then back on the 21st. It was a most peculiar sensation, taking off from the sea — you thought the thing would never get off. And the noise! 'It was very good, though, once you climbed into the air.'

Arthur Robert was talking about travelling on the slow, two-engined Cloud of Iona, one of the brightest amphibians in Jersey Airways' fleet and a biplane which he used when the sea was the only airport in Jersey. He was remembering a time when it was genuinely believed that a calm sea was the best thing for passengers since man discovered the horse.

In the 'Weekly Post' for July 22, 1927, a reporter's account of travelling the skies is given. Parts of the account make entertaining reading: 'An Avro biplane (Avro 536 — G-Boy) fitted with a 130 hp Clerget engine is now in the Island and is giving passengers flights.

'The petrol used is Pratt's No 1 Spirit and the lubricating oil is pure castor oil. The machine belongs to the Surrey Flying Services of Croydon and is piloted by Captain S F Woods, a pilot who served on various fronts in the RFC and during the war.

'The sensation of actually flying seemed, to our representative, to be little different from motoring at a high rate of speed. One is in the air before one actually realises the fact, but the main sensation once aloft is the complete stability of the machine.

'After circling the town (at about 1,500 feet) the machine headed through the centre of the Island, passing over St Lawrence and St Peter to the racecourse and the Militia Camp. It did not go directly over the camp for fear of frightening the horses.

'The machine headed back to its temporary home at Bel Royal, passing through a cloud on the way. Circling over the landing place once or twice the machine came easily to rest on the sand. We would like to point out that before the flight the engine controls, rudder, etc, are thoroughly tested.'

The comparison between the Avro biplane and a motor car was very apt at the time, because the first planes to the Island (including Benoist's Sanchez-Besa biplane of 1912, which averaged 60 mph) were only marginally faster than some of the quicker two-door coupés.

October, 1930: England's largest amphibian, the Saro Cloud, beached at West Park

February 7, 1931: D V Ivan of the RAF has his plane towed across the Esplanade in search of petrol

Who will Fly to the Islands?

ON Tuesday, May 18, 1926, the Jersey General Engineering Company received a telegram from Imperial Airways which said that a seaplane would be arriving the next day with room for four passengers to Guernsey, two to England.

No one seemed interested in taking up the offer, but a large crowd congregated on the quayside at noon on Wednesday hoping to see the plane arrive. Sadly, with no passengers to collect, it never came, which begged the question — who would travel by air?

It was a question which had been puzzling the British Marine Air Navigation Company for some time, ever since they had initiated a regular Guernsey — Woolston service (out of season) on September 1, 1923. This was the first commercial flying venture between the Channel Islands and the UK, and made St Peter Port the islands' first commercial airport, and one of the first seadromes in the world.

However, it was difficult to make money on the route and even when BMANC amalgamated with Handley Page Transport Ltd, Instone Air Line Ltd, and Daimler Hire Ltd, in 1924, to form Imperial Airways, the Southampton-Guernsey route was only maintained on an ad hoc basis, grudgingly, to satisfy British Government requirements.

According to these requirements, the government had agreed in principle to pay Imperial Airways £10,000 for a UK—Cherbourg—Channel Islands service and a further £21,000 if the company's planes flew a minimum of 60,000 miles, in total, on established routes each year.

The subsidies were tempting but the Channel Islands' route was not as rewarding as other less seasonal destinations. The Cherbourg connection fizzled out almost immediately (the company was unable to use the French port as a seadrome) and although 80 mph Supermarine Sea Eagle flying boats made several crossings to Guernsey, their flights were infrequent and were made (according to some sceptics) only because they were necessary if Imperial Airways were to receive their money from the British government.

This doesn't mean that Imperial Airways were uninterested in the Channel Islands as a potential money-making route. They tried to arrange sea crossings between Jersey and Guernsey so that passengers could

1933: The first autogyro to visit Jersey on St Aubin's beach

connect with their seaplane before flying on to England, and they made plans for a Guernsey-Southampton -London service, as well as planning a Guernsey—Paris route (a combination of air and rail).

All of these schemes had merit, but who would pay, for example, £3 18 shillings for a one-way ticket to Southampton when a return trip by boat cost a great deal less? And although the tourist trade had potential (on one Bank Holiday Saturday in August, 1924, over 3,000 passengers travelled to and from Jersey), why pay as much for your flight when a return fare PLUS accommodation, by sea, would cost you a great deal less? Until planes grew bigger and fares went down it was best to leave air travel between the mainland and the islands as an idea . . . a good idea perhaps, but something to be left to braver men.

Men like the members of the Jersey Chamber of Commerce, for example.

Since 1919, when Lloyds Weekly News was brought to the islands by air during a national rail strike, the council of the Jersey Incorporated Chamber of Commerce had been watching the development of air communications with interest.

The council, and in particular W G Bellingham (president) and E F Guiton (secretary) realised the potential of air transport and considered it from an islander's point of view. They were interested in the long-term possibilities, not the short-term 'grab the money and run' philosophy of unscrupulous outsiders.

And, they were interested in all forms of cargo. Passengers were important, but so, too, was freight — including mail and the export of fresh farm produce to mainland markets.

In 1920 they had written many letters to the British Government and to air companies, in the hope that someone like Instone Air Line Ltd would operate a Channel Islands/UK/France route for the carrying of mails (Instone already ran such a route between London and Paris).

They also prepared a petition which was sent to the then-Postmaster General, F G Kellaway and to the States. The petition included the following: 'To the Rt Hon Frederick George Kellaway, MP, His Majesty's Post Master General. . . . That the situation of the Channel Islands is admirably adapted to the use of a service of amphibian machines from the Thames in London, the ports of Jersey and Guernsey being eminently suitable for such machines.

'That there is a great demand not only for a more rapid service of letters, but also for speedier conveyance to the islands of newspapers and foodstuffs such as fish, also for transport to England of flowers and fresh perishable fruit. Under present conditions newspapers are at least one day old on arrival; the delay in transit is also very detrimental to flowers and certain foodstuffs of a nature to deteriorate rapidly.

'That there is little doubt that such a service would receive adequate support and that it could later be supplemented by an addition for the transport of passengers.
'24th May, 1921.'

When the Post-Master General replied, it was with a vague half-promise that if a private company was prepared to fly regularly to the Island, he would consider letting it take the mail — but the Post Office weren't going to be persuaded that easily. After all, railway companies had been providing an adequate postal service for years and would continue to do so, even if Jersey was more than 100 miles away from any UK railway track.

Guiton and Bellingham continued their campaign. The possibility of an airship was considered (an airship was then being built which could carry seven tons of produce) and Guiton also visited the Vickers works in London to study new amphibian designs with the intention of running a direct air link between St Helier and the Thames.

Despite the frustration of receiving polite conversation but no promises from people with money, and planes, the local Chamber of Commerce persisted and letters were sent to Chambers of Commerce in all of the major UK cities asking that pressure be placed on Government and city councils in the hope that sooner or later a regular, Jersey-mainland air link would be estab-

The St Aubin's Bay was one of Jersey Airways' regular aircraft on the Jersey route

lished. However, eight years later the problem was as bad as ever. There had been a glimmer of hope during those years that a direct flying service would be introduced, especially in August 1928, when Imperial Airways announced they would introduce their Supermarine Sea Eagle flying boats on to the Jersey-Southampton route — but within a year all hopes had been dashed.

In early 1929, Sir Eric Geddes, chairman of Imperial Airways, announced that poor flying weather, delays at air terminals, the shortness of domestic routes and well-established rail services all contributed towards making domestic flying unprofitable.

Confidence in any long-term plans for the Jersey route was also dented on August 8, 1929, when an Imperial Airways Supermarine Seagull IV (GAAIZ), in attempting to take off from St Aubin's Bay on a sunny day and across a calm sea, hit a rock, tore away the lower part of the fuselage, and had to be shipped across to Southampton aboard the ss Haslemere.

Imperial Airways began to invest their time and money in longer, more lucrative routes, including the rich air path to India. This meant that they were leaving the way open for other air companies to step into their shoes as they turned their back on Britain, but this was of little comfort to Jersey and to its Chamber of Commerce.

Lack of quick communication between the islands and the mainland also annoyed one or two British people, including Unionist MP William Lindsay Everard:

'I think I am right in saying that today there is no service at all and very little likelihood, unless a subsidy is forthcoming, of this service starting again . . . It is a very poor lookout for those outlying parts of the country when they cannot be connected with air communication . . .' (W L Everard, House of Commons, May 13, 1930).

But still Imperial Airways continued to run down their local services, which meant that the Jersey Chamber of Commerce had to think again. They already had one idea in the pipeline, an ambitious dream which, if it became reality, would alter the island's way of life forever.

What Jersey needed, it would appear, was something like a fly trap to attract a major airline company to the Island and to keep it here. What was needed most was an airport . . .

Men of Vision

ON June 4, 1930, the Jersey Chamber of Commerce discussed the implications that an airport would have for the island. A letter was then sent to the Air Ministry and an invitation extended to Sir Alan Cobham to visit Jersey to consider whether an airport was a viable proposition. He didn't come over — but on Monday, October 17, 1932, Mr L Balfour did.

In the intervening two years a great deal of correspondence had crossed the seas concerning the need for a Jersey airport, and Balfour's visit was the result of several Chamber of Commerce meetings at which the actual positioning of it had been discussed.

Several sites were considered, including the race course at Les Quennevais, the old prisoner of war camp at Blanches Banques, Les Landes, Les Platons, Mont La Mare and a site near St Peter's Barracks. As a representative of the Portsmouth, Southsea and Isle of Wight Aviation Company, Balfour was interested in establishing an airport for his own company — which would thus lead to regular flights between the Channel Islands and Portsmouth. He eventually narrowed the choices to two — Mont La Mare (St Ouen) and the St Peter site — and on the third day of his visit, he went again to consider the two before submitting, a month later, a lengthy report on their suitability. By that time the Jersey Chamber of Commerce had pushed even further ahead with their idea.

John Le Marquand, president of the Jersey Chamber, had already seen the officer in charge of government lands, Captain Vernon Cooke, about the possibility of acquiring land for the airport.

On November 28, 1932, he saw the Lieutenant-Governor, Major-General Willis, with Balfour's recommendations. The Lieutenant-Governor was helpful and keen, and the next step was to take Balfour's recommendations (which were now largely those of the Chamber of Commerce) to St Peter to find out what the reaction would be from the parishioners and the Constable.

A meeting was arranged at St Peter's Parish Hall on December 8, when members

A police guard for one of the entries in the 1933 air display

of the special com-
mittee of the
Chamber of
Commerce met the
Constable, Mr John
du Val, Mr Harold
Benest (surveyor) and
a number of residents
who owned land in
the area. Mr Benest
was asked to prepare
a survey of land
intended for use as
an airport, a small
sub-committee under
the chairmanship of
the Constable was
formed to find out
which land would be
needed and who
owned it, and the Air
Ministry were again
informed of the

Mrs Victor Bruch (right), the first woman to fly solo around the world

progress that was being made. The scheme
was growing each day and it couldn't be
long before it would have to be placed
before the States, who would have to have
the final say in the matter.

On the surface at least, everything had
gone remarkably smoothly, if very slowly in
the initial stages.

The Chamber of Commerce's initiative
and more than two years of letter writing
and consultation were proving fruitful.
Even the Lieutenant-Governor had taken a
keen interest, and his negotiations with the
War Office in London smoothed the way to
buying land in Jersey, owned by the
Crown, on condition that if the barracks at
St Peter were pulled down, they would be
rebuilt elsewhere. If it was impossible to
rebuild them, a separate guarantee was giv-
en that any troops stationed in the Island
would be housed at Jersey's expense.

British Government officials did voice
some concern about the project — the
Portsmouth, Southsea and Isle of Wight
Company were not on the Ministry's civil
aviation list — but in Jersey a far greater
obstacle needed to be overcome.

Would the States agree to an airport

being built on Island land?

Carteret presented plans to the States
outlining the area of St Peter which would
be needed, plus details of the scheme
researched so diligently by the Jersey
Chamber of Commerce. There was a cover-
ing note: 'The Executive Council (of the
Chamber of Commerce) . . . has always
intended to submit the results to you,

in the hope that the establishment of an
island airport might become a States project, rather than one of private ownership.'
That note was needed. In 1933 no private
company would have been given permission to buy land and use it to house and to
run planes along, and Jurat Reginald Malet
de Carteret, president of the Finance
Committee, had already scathingly
described the airport project as a 'wildcat
scheme'.

The States referred the project to the
Piers and Harbours Committee and to the
Finance Committee for their consideration.
It would appear that there would be a lot of
nail-biting before the first bulldozer moved
across St Peter's land.

Thankfully, once de Carteret had recov-
ered from the shock, he treated the whole

thing with commendable objectivity and arranged a meeting in June with appointed representatives from the Chamber of Commerce plus Balfour and other agents from the Portsmouth, Southsea and Isle of Wight Aviation Company. More meetings followed and the Air Ministry was again consulted until, eventually, the States engineer was directed to prepare a plan of the intended site. Meanwhile, the Attorney-General was instructed to prepare a draft law authorising the expropriation of any private land which might be required.

But there were still headaches — and Deputy Edward Le Quesne was extremely worried that they would be never-ending. What, he wanted to know, would be the long-term implications of owning an airport when some UK privately-owned airfields were already going bankrupt?

Would an airport be a perpetual burden to the Jersey taxpayer and how could anyone say with any confidence that aeroplanes had an assured future? What about helicopters?

They were reasonable questions and it was left to the Air Ministry to provide at least one answer. 'No,' the States were assured by a letter on October 10, 'fixed-wing aircraft are here to stay'.

There would be many problems to come, of course, and many people weren't entirely convinced about the need for an airport or noisy aeroplanes landing in their parish, but it would appear that a States owned airport had changed from a possibility to a probability — and it would be the Piers and Harbours Committee assisted by the Finance Committee, who would end 1933 with a Jersey Airport bill ready to be put before the States, thereby putting the finishing touches to work initiated by the Jersey Chamber of Commerce so many years before.

In 1933 during an air display at La Moye one of the biplanes gave a foretaste of what was to come only a few years later by dropping a bomb — accurately — on a wooden shed and destroying it

Meanwhile, on the Beach

IN November, 1938, the Jersey Weekly Post was able to boast that Jersey owned the second busiest airport in the UK. In all, 34,559 passengers had walked across the airstrip and only Croydon Airport was busier.

It would appear that this remarkable statistic justified the Jersey Chamber of Commerce's persistence, although their efforts might have gone unrewarded if three local businessmen hadn't risked a lot of money and a great deal of time in proving that people would fly to the Island, and would pay to do so, even if the Jersey centre of operations was a sandy beach, rendered inoperative at high tide.

After Imperial Airways had dithered about, trying to justify a route between the UK and the Channel Islands, and doing no-one much good in the process, more than one company tried, unsuccessfully, to establish a regular Channel Island-UK flight schedule.

The Tours and Travel Association, for example, with six-seater Supermarine Seagull amphibian planes (equipped with Napier Lion engines and fitted with Handley-Page slotted wings which could be easily folded) operated intermittently during the summer of 1929.

Eventually, bad weather, underfunding and accident caused the venture to fail, although the chief pilot with the association, S S Kirsten, believed that the Channel Island route did have potential.

In 1930 he and R B Mace came to Jersey, determined to make air travel pay. They owned one plane — Saro A.17 (G-AAIP), a large, all-metal amphibian flying boat with retractable undercarriage and powered by two 150hp ADC Hermes 1 engines. Called the 'Silver Bat' she was able to take four passengers with luggage, six without.

For £6 10s. (£6.50) it was possible to make a return journey to Woolston airport (Southampton). Tickets were available from Bellingham's, Mulcaster Street, and, if you wanted to charter the plane for a visit to France, or to spend a little time flying above the Island, Kirsten and Mace were happy to put their plane at your disposal.

Theirs was a bold venture, but it failed. Although they were prepared to be flexible and would fly anyone anywhere (if the price was right), too few bookings on a small, four-seater plane would never earn

Jersey Airways' mobile West Park air terminal

1937: A line-up of Jersey Airways' crew

Passengers wander around the aircraft — a far cry from today's world of high-security airports

A police guard for a Jersey Airways DH 86

them a great deal of money, especially when the tourist season ended and the Jersey winter began in earnest. Kirsten and Mace loaded up their plane, headed north and eventually ran air services between Blackpool, the Isle of Man and Liverpool. On the face of it, at least, another Jersey flying venture had failed to get off the ground.

However, although the two men had not earned a great deal of money for their work, their enterprise hadn't passed unnoticed. Unlike Imperial Airways, who wanted to operate INTO the Island, without paying more than a fleeting visit to pick up passengers and freight, Kirsten and Mace made Jersey their base and, if they hadn't done so, businessmen W L Thurgood, L T H Greif and Jersey farmer J A Perée might not have launched one of the most successful, 'home-grown' airlines — Jersey Airways Ltd.

Thurgood had owned the People's Motor Services in Ware, Hertfordshire, before selling it and moving to the Island and, as he settled into the Jersey way of life, he noticed how wide and flat St Aubin's Bay was and how cheap it might be to run a plane service from the Island's shores.

With his two friends and an initial investment of £120,000 he launched Jersey Airways in 1933, the company's first flight leaving Jersey for Portsmouth on December 18 that year. His 'airport' was a square half

mile with the centre 440 yards due south of the second tower in St Aubin's Bay (the Receiver-General, J F Giffard, who defined the airport, hadn't got it quite right because it should have been the first tower) and this became one of the only authorised airports in the world to be submerged twice daily and to have a timetable controlled by the tides.

Jersey Airways could cope with these problems. The company was based in the Island and, by keeping costs as low as possible and by continually investing in newer, more reliable planes, it made a profit. After a year of operating eight-seater De Havilland DH 84 Dragons, the company switched to 14-seater De Havilland DH 86 Expresses. A half-empty DH 86 was more economical to run than the full, four-seater 'Silver Bat' that Mace and Kirsten had been trying to run from the Island a few years earlier.

Despite being formed midway through the winter of 1933, during the first year of operations 20,000 passengers were ferried between the Channel Islands and the UK and 96 per cent of all scheduled flights were made. Regularity encouraged people to fly. Islanders (and tourists) were starting to accept the idea of flight in a way that their parents and grandparents had come to accept the motor car and train. In 1935,

1934: A Ford Tri-Motor was called in to help ferry passengers during the Whitsun rush

despite profits of only £361, Jersey Airways flew 450,000 miles and carried 24,717 passengers as well as 152,764 1b of freight — slowly but surely the Island was becoming dependent on the airline, and Thurgood must have been delighted that his company was making headway when so many others had failed.

Southampton became the main UK terminal, a regular Jersey to London flight was introduced, and Jersey Airways had high hopes that a twice-weekly Paris to Jersey flight would prove successful (it wasn't and neither Air France nor Jersey Airways could make this early French connection pay). The optimism, growth and reliability of the Island-based air service was a powerful, practical argument in the Chamber of Commerce's bid for a purpose-built airport.

Cautiously the Island was coming round to the Chamber's way of thinking but, if the beach was suitable for at least 96 per cent of all flights across the water, why build on land? And why not stick with amphibious planes, which could land anywhere, be it grass, sand or water?

If all diehard reactionaries wanted further proof that the sea and all it controlled wasn't to be trusted, that proof came on July 31, 1936, when the Jersey Airway's amphibian, Saro A. 19, Cloud of Iona (G-ABXW) took off from St Peter Port (which had no landing strip) shortly after 7 pm, bound for Jersey. She never arrived. No-one knows exactly what happened to her, although she landed safely in the sea somewhere south of Jersey before tragedy struck.

Visibility was poor before she took off and when she didn't report as usual, the Jersey traffic manager of Jersey Airways, Mr Wieland, put out an SOS. A week later the eight passengers and two crewman were washed up, dead, on the French coast.

If the craft had landed at a properly equipped, well-lit airport and not on water, which held her tight to the waves, ten lives might not have been lost. But some good came from the tragedy because the loss of plane and passengers helped to convince some of the remaining sceptics that an airport at St Peter made sense and, by now, the airport had got past the planning stage and was in the process of being built.

This undoubtedly pleased Jersey Airways, and also the Great Western Railway and Southern Railway who, in 1937, bought shares in the company on the assumption that 'if you can't beat them, join them'. They knew that air travel was here to stay and air/rail tickets, in combination, made a tempting package for the many thousands of people living in Britain who might enjoy their summer holiday on a sunny, holiday Island.

The Red Arrows have appeared each year since 1965

Honour the Few

DURING the morning of May 26, 1938, the RAF in the form of a Walrus, Swordfish, Nimrod, Osprey, Wildebeeste, six Ansons and three small RAF flying boats buzzed the Island as celebrations to mark Empire Day reached a climax. That afternoon Islanders were invited to the airport where, for a nominal charge, they could take a closer look at the aircraft on the ground. The proceeds went to the RAF Benevolent Fund.

Today, well over half a century later, and RAF planes still buzz the Island at least once a year (with the authorities' permission), normally on the second Thursday in September when the afternoon air display is the highlight of Battle of Britain week.

The organisers of the oldest Royal Air Forces Association in existence, Jersey Branch No 487 (founded by ex-members of the RFC and RAF in 1925) have continued to collect for charities associated with the RAF and half a million pounds has been raised for the RAFA Benevolent Fund and the RAFA Convalescent Home at Sussexdown since 1953.

It was in that year that the first official Battle of Britain Air Display was held in Jersey, when the new Meteor jet aircraft made their first appearance accompanied by the actor, Trevor Howard, who autographed programmes on the wing of one of the aircraft at the airport.

As the years have passed, the fly-past has become one of the main events in the Jersey calendar although the airborne stars of the display have varied from one generation to another.

On Thursday, September 13, 1956, for example, the Canberra jet bomber with a wingspan of 63 feet was the star of the show.

By the early 1960s the famed Red Pelican team, six brilliant red Provost jets, were stealing the show and then, from 1965 onwards, in the slipstream of the Red Pelicans came the Red Arrows, formerly the Black Arrows, a team of pilots from the Central Flying School at Little Rissington, Gloucester, whose nine scarlet Gnat or Hawk trainers have been welcome guests at the display ever since.

Building an Airport

WHEN the Jersey Chamber of Commerce submitted their airport plans to the Piers and Harbours Committee, one of the most important factors that had to be taken into account was cost.

It had been reckoned that the area needed for the airport was about 70 fields (145 vergées or 64 acres). After every land owner had been contacted for a second time, it was reckoned that land could be bought at between £80 and £140 per vergée. In all, the Chamber of Commerce were told, the 45 vergées would cost £12,987 5s.

A round-up of the costs became: Land, £14,000; Surfacing, £3,000; Hangar (for one medium-sized airliner) £2,000 — an initial, optimistic forecast of £19,000.

Eventually, the Chamber of Commerce revised their estimate to £30,000 and they, and the States discussed ways of getting that money back. Should (for example) the airport be leased out to someone like the Portsmouth, Southsea and Isle of Wight Company?

Perhaps — but as the debate continued, new and ever-worrying problems dogged the project. What about Customs, ambulance services, fire engines, warning lights and adequate roads leading to the Airport? Would the War Office give their land and bar racks, cheaply, if required? And what would happen in the future if the airport wasn't big enough?

As Jersey Airways continued to do well, this last problem became increasingly worrying, and it was a very cautious Jurat Malet de Carteret who chaired a joint committee meeting which arrived at the idea of splitting the land into three. There was the land to be used, land to be monitored to ensure that no obstacles (ie buildings) would, in the future, be erected there without consent, and War Office land.

On April 3, 1934, the joint committee's proposals were placed before the States and accepted. In presenting the Bill, Jurat G F B de Gruchy, president of the Piers and Harbours Committee, told the Island government that if they wanted an airport they must act quickly and, to their credit, most States Members, realising how important air travel had become, agreed.

A cautious note, however, came from Deputy Le Quesne, who suggested that they should buy the land but not build on it (someone else could do that) and, even after the States had committed themselves to a Jersey airport, his fears that they were over-reaching themselves continued.

As the concept of a Jersey airport shaped itself into fact, Le Quesne's warning that the Island was stumbling into a horrendous, bottomless pit of financial worries seemed justified.

In February, 1935, the Piers and Harbours Committee were given another £2,200 so that another 14 vergées of land could be bought.

In October, £700 was voted for a wireless service and in the same month £17,000 (in the form of a loan) was requested so that building work could continue. No one, the States were told, would tender for the contract at the estimates given.

The next sum, £27,000, for continuing the work and for flood-lighting, was requested in November, 1935, and a year later, the problem of night-flying equipment again came before the States. An extra £7,000 was needed (perhaps more?) and as Jurat de Gruchy went cap in hand to his fellow States Members yet again, it was against a backdrop of increased hostility.

The local newspaper, the 'Evening Post', was especially biting in its criticism of a State-owned airport and de Gruchy's broad frame was an ideal target to attack.

In hindsight they, and the other anti-airport lobbyists, were doing Jersey a disservice but no one realised this, not even de Gruchy, whose comments that 'lights would only be switched on when news was received of approaching aircraft' (thereby saving the Island money) seems strangely naive in the late 20th century when so many flights are made at night, but he had to justify his continual requests for more money in the face of such criticisms as:

Contractors Ernest Farley lined up and ready to build an airport

'The place (the airport) is too big and the States (are) not studying economy,' quoted by Deputy Le Quesne.

By February, 1936, another loan of £30,000 was requested. 'Certain items,' the States were told, 'had not been contemplated'. However, de Gruchy felt that this, plus the £97,000 already voted, would be sufficient to complete the airport (which was true) although he had to return a few months later to ask for more money for a direction finder and another 21 vergées of land to put it on.

By now the airport was nearly finished and ready to take air traffic, although its eventual cost had been in the region of £127,000 (the overall, estimated TOTAL expenditure for all States' projects and salaries in 1936 had only been £391,829) and there was the added financial burden of maintaining it.

As the Bailiff's wife, Mrs A M Coutanche, officially opened Jersey Airport on March 10, 1937, it did not pass without censure in the press.

The 'Evening Post' launched another attack on this costly 'albatross' and took the States to task for refusing to allow anyone but a few carefully-chosen guests to the opening ceremony, pointing out that the airport had been paid for by all islanders, not just the States.

But at least it had been built. That it stood at all was the Chamber of Commerce's and Jurat de Gruchy's triumph and, as the president of the Piers and Harbours Committee stood on the long, 980 yard runway on that March afternoon, did he have any real idea of its importance to Jersey? He would have been able to glimpse the future more keenly than Miss Gertrude Bacon, but wouldn't have dreamt that jet aircraft, 200-seater airbuses and a million and a half passengers a year would be flying into the Island in the 1990s. His dreams would have been of a more immediate nature, and would have included the problem of paying back the £100,000 loan which had been necessary to complete his airport.

Eventually that amount was repaid and, within a year, an estimated 20,000 visitors had flown to Jersey, many of them holiday-makers from the mainland. Whether they justified de Gruchy's claims that 'an airport is a good investment because it will induce a better class of visitor here' is another matter, but as all types of cargo, human or otherwise, began to arrive with even greater frequency, an occasional wry smile must have crossed de Gruchy's face. More people and more freight meant more wealth. If there hadn't been an airport, the island would have suffered. It was a gateway to prosperity.

Original dimensions

WHEN Jersey Airport was first built, in 1937, it was based on 218 vergées of land (97 acres), the landing area alone being 77 acres. There were four flightways which provided potential landings in eight directions. These were:

North—South (258 yards).
North-east—South-west (720 yards).
South-east—North-west (720 yards).
East—West (980 yards with a white concrete centre line).

The building work had cost £98,000 (E G Farley, builder) and the initial, modest plans for one hangar (220 feet by 100 feet) had been altered to allow a second hangar to be built. At approximately every 90 yards around the airport boundary were warning lights; four floodlights, with 11/4 million candlepower each, were positioned so that a pilot could land against any wind direction, without being blinded.

Red obstruction lights were placed at the tops of high buildings in the area (including St Peter's Church) and other safety aspects included a look-out, using an Aldis lamp with red, green and white screens, a radio and control tower, a Crows Nest for a look-out man in contact with visiting planes by radio (control officers were initially in contact with aircraft by radio W/T-Morse code. This was later replaced by R/T-speech) and a steel tower beacon at Red Houses shone as 'the lighthouse of the air'.

The airport itself stretched from the Don Bridge Road to the Barracks and to the edge of a côtil bordering Jubilee Hill. A small road to the Barracks from the main Don Bridge road had been removed and the whole area had been levelled and grassed. In the centre of this green sward was the word 'Jersey' in large white letters, and in a circle — this was easily visible from the air.

The terminal building had been designed by architect Graham Dawbarn. The fourth floor housed the control and wireless rooms, the third floor

A final coat of paint before the opening ceremony

was for the meteorological men, the second floor contained a restaurant and terraces. Access to the 'ample space of the first floor promenade' was obtained direct from the forecourt where deck chairs 'together with the services of a snack counter' were provided. The main space in the building, however, was at ground level, and extended to either side of the main control block 'in the same manner as the wings of a swallow in flight' so that in the future, if necessary, extensions could be added. On the ground floor was the high-ceilinged main hall. To the right, as you entered the main building, was the incoming section of the airport (to the left, departures).

Space had also been provided for a mails room which meant that at long last the post would be flown to Jersey on a regular, daily basis.

Left: Lady Coutanche performs the official opening ceremony

Below: Guests mingle with passengers as the first flights arrive

1937: The airport looked very different than today with its grassy runways and rural outlook

1938: An aircraft undergoes an overhaul

Worrying Losses

IF Jersey Airways' blossoming success, as they continued to make two flights a day from their sandy airport at West Park, was one of the factors which persuaded the States to continue their expensive building programme at St Peter, perhaps they also thought that the airport would be theirs to use, as they saw fit, once it was finished.

This was not to be. They DID have a 14-year lease with the States for the main hangar, plus office space, at a cost of £1,800 a year (slowly but surely Jurat de Gruchy was determined to get his money back); they WERE given a provisional concession to bring daily newspapers into the Island after lengthy negotiations with Fleet Street proprietors (within a year this was given on a long-term contract); and, from June 1, 1937, the Post Office DID allow them to fly the mail into the Island, BUT it is interesting to note who else had been invited to that opening ceremony on March 10, 1937.

As well as representatives from Jersey Airways, there was Mr P Le Masurier, Jersey agent for Imperial Airways, and Mr Roberts, vice-chairman of British Airways. Both men must have looked at this new, sophisticated (as it was then) airport with interest.

Jersey Airways might have held a monopoly in 1937, but who knew what the future had in store?

In 1937, 33,314 people and a massive 1,090,782 lb of freight were flown in or out of the island by Jersey Airways. The company were planning new routes (to Exeter and to Dinard) and, on the surface, it seemed that they were doing exceptionally well. However, when the audit books for 1937 were passed it was discovered that more than £10,000 had been lost in the year's operations.

The obvious question was 'why?' and, not surprisingly, the board looked carefully at their books before waging a bitter war against excessive overheads which, they claimed, had no right to be there. A prime target for their attacks was the excessive cost of operating out of Jersey and, as the September edition of 'Popular Flying' explained in 1937, they weren't the only ones who had to pay dearly for the right to use the airport.

'Pilots contemplating making use of Jersey Airport are advised to make themselves familiar before arrival with the scale of charges enforced there, otherwise they are likely to receive a rude shock upon receipt of the bill.

Prices include: 'For every person (including pilots and crew) 1s. 3d. per landing, 1s. 3d. per take-off, 2s per aircraft landing or take-off). Hangar space, 3d. per square foot floor-space taken up, per month.'

If the charges had been too excessive, Jersey Airways would have passed them on to their passengers (but they were reluctant to do so) and despite

Airport Controller C P V Roche

Air traffic control in the early years

Having disembarked, passengers collect their luggage

March 10, 1937: Jersey Airport viewed from across the apron

their complaints, at least their planes had a proper Jersey base and hangar accommodation. Before Jersey Airport was built, aircraft would be flown back to Heston or Portsmouth for the night. And Jersey Airways had a large, ever-growing airfleet which by 1939 was worth a great deal. They were keen to expand, to open up new routes and to buy bigger and faster planes. The 14-seater De Havilland Express, with a maximum speed of 170 mph (a far cry from those pioneer days of Kirsten and Mace) was introduced on to the Island route and, as the shipping companies found that the number of passengers wanting to travel to Jersey by sea had been drastically reduced, so the number of passengers keen to fly to the Island increased dramatically.

In May 1939, the prototype of the De Havilland DH95 Flamingo monoplane (G-AFUE), an 18-seater of advanced design and capable of 239 mph, was loaned to Jersey Airways, who were confident that they would continue to be the Island's number one air company for many years to come. They had good cause to look hungrily forward. After approaching the States of Guernsey they had been given permission to build their own airport at La Villaize (opened on May 12, 1939 at a cost of £114,500) and despite war fears, they were petitioning Jersey's Piers and Harbours' Committee, urging that an extension to the north-south runway at the airport would be to everyone's advantage.

It would have been. Most islanders appreciated their airport by now, despite a horrendous air crash on November 11, 1938, when 14 people were killed, and despite the occasional swipes at the airport and all it represented by the 'Evening Post': 'As we have contended all along, Jersey Airport will, in the end, not only be a costly luxury to the Island at a time of peace, but a potential danger in times of war.' (*'EP', April 2, 1938*)

Channel Islanders are, by nature, a cau-

Jersey Airport from the forecourt when it had only just opened

tious race and the 'EP' was right to keep a watchful eye on what the next developments would be at St Peter. However, they, and all of the islanders, were now enjoying an increase in mail (which was flown in daily), speedy access to markets, which was invaluable for growers, and an improved air-sea rescue service co-ordinated by Air Traffic Control.

An airport, properly equipped and well run, could only benefit an island race, but no one knew what the future held in store as Chamberlain tottered back from Munich after his historic meeting with Hitler in September 1938, and 1939 dawned.

May, 1939: A de Havilland Flamingo over Gorey

The War Years

A 'POTENTIAL danger in times of war' is how the 'Evening Post' had described the airport in 1938. It was not to be that dangerous — Guernsey airport, only recently built, was to see more action — but no one, as yet, knew this.

Instead, when civil aviation services were suspended at the end of 1939, the airport was gradually cleared of planes which were needed in Britain.

Pilots and ground staff were also needed, leaving the airport a shadow of its former self. Like 11 other civil airports, it was requisitioned and placed in the hands of the military.

For a year it lay in a kind of limbo as the 'phoney war' of 1939/early 1940 allowed islanders and the British an uneasy peace.

Restrictions were relaxed — it was quickly realised that the airport was too important to close, and air links continued between Jersey, Guernsey, Shoreham and London, for limited freight and passenger services, until German occupation of the Channel Islands became inevitable. In March, 1940, there was an Easter rush of UK holidaymakers who flew to Jersey determined to enjoy what turned out to be their last Channel Island holiday for quite some time. But, even as they relaxed on the beaches, the German army was making huge strides across Europe. By May, it was only a matter of time before they arrived on the French coast and if they reached Cherbourg or St Malo, the next logical stepping stone to Britain was the Channel Islands.

As one nation advanced and another retreated, there was a great deal of activity at Jersey Airport and June, 1940, was particularly busy. On the night of June 11/12, for example, 36 Armstrong Whitworth Whitley bombers refuelled there on their way to carrying out the first RAF raid in northern Italy. The RAF squadrons involved were numbers 10, 51, 58, 77 and 102 and the target was the Flat works at Turin. Only 11 aircraft reached the city.

Three days later the Air Ministry ordered airlines flying into the Channel Islands to

Oberleutnant Richard Kern meets the Bailiff and Attorney-General

A USAAF P47 Thunderbolt which crashed in Jersey. Pilot Walter Davis was taken prisoner

suspend operations and advised the Bailiff and Lieutenant-Governor that the Island should be evacuated. Wing Commander J D Miller arrived at the airport the same day and for four days ran a staging post for many of the aircraft getting out of France, one of the distinguished arrivals from France being General Charles de Gaulle, who paid a fleeting visit on June 17th.

During this month about 50 members of the Number 23 Wing Servicing Unit were located in the Island to help to service and refuel military aircraft passing through, but by June 20 their work was coming to an end. On that day all Wehrmacht units in France received the following message from Berlin: 'Occupation of British Channel Islands urgent and important.'

Jersey continued to demilitarise as quickly as it could. All aircraft (apart from one) were flown to Heston. Anything which could be used by the Germans but couldn't be flown out of the Island was destroyed.

Between June 19 and 21, nearly 400 passengers were flown to the mainland before the Jersey authorities stopped evacuation by air as it was revealed that evacuation by boat would be possible.

As well as people, more than 6,000 lb of aircraft stores were sent to Heston, a Scottish Airways DH Rapide was flown out with a valuable spare engine fastened to the wing, and all that was left behind was a DH86 Express (G-ADVK), then undergoing major overhaul (the RAF rendered it useless), and petrol storage tanks connected to tap water. At a moment's notice they could be swamped and the petrol made useless.

The airport was, to all intents and purposes, dead. All that remained was a handful of Jersey residents who had volunteered to man the place, knowing that the next planes to fly in would be German. They were in command as the islands were overflown and photographed by reconnaissance planes from the V111 Air Corps during the last week of June, a particularly tragic week for the Island for, on June 8, six Heinkel He 111 bombers dropped their bombs at La Rocque and St Helier, killing eleven civilians and wounding many more.

On July 1, messages in pouches were dropped telling the Island authorities to announce their surrender by placing large

July, 1940: The first occupying forces arrive in a Junkers JU 52

white crosses in three key areas, including the airfield. As this was being done, the crew of a Dornier Do. 17z from 2 Staffel Aufklaerungsgruppe (F) 123 noticed that the surrender signs were being unravelled and, while Lieutenant Zuleeg circled as cover, touched down in Jersey.

So 25-year-old Oberleutnant Richard Kern was the first German to 'capture' the Island and in the process to take the Airport. He first met Airport Controller Charles Roche and then asked to see the Bailiff, Alexander Coutanche. By 5 o'clock, surrender terms had been agreed and, within a short space of time, eight Dornier Do. 17z aircraft and two Junker JU 52/3m troop transport planes had arrived, and so, too, had nine officers, one medical officer, one public servant and 51 ncos and enlisted men. The Island was well and truly occupied.

For the rest of the summer and into the early part of 1941 Jersey Airport was busy again. With the help of the Reichs Arbeits Dienst and Luftwaffe ground staff, eight wooden hangars were built to house the German aircraft (seven were dismantled and taken to France in May 1944, the other eventually became the property of Hallmark Hire Cars), concrete taxiways were laid between these buildings and the airfield proper and a test range for aligning aircraft guns was built nearby.

The aerodrome, as the Germans called it, was placed out of bounds and islanders living nearby were moved as it became the largest exclusion zone in Jersey. On September 19, 1940, the parish assembly at St Peter were told to approve plans for a new road to the airport (which they did) and it was anticipated that a road would eventually be built to run at an angle to Beaumont Hill and to the Luftwaffe ammunition tunnels, to make the reloading of planes faster and easier.

There were grandiose plans to turn both Jersey's and Guernsey's airports into bristling, armed camps, from which regular sorties could be made against the British. After Hitler had ordered that the Channel

Islands be turned into an 'impregnable fortress' in 1941, it was believed that three million tons of building material would be needed at the airport, to reinforce it, and there were rumours that a large squadron of Junkers Ju 88s, plus 2,000 extra personnel, would be drafted in.

But, by February 1941 the war had moved away from the Channel Islands, and for the next three years the airport was of minor consideration to islanders and occupying forces alike.

When the Germans occupied the Channel Islands in 1940, they fully expected the Allies to make some sort of attempt to reclaim them. They also believed that they would be useful bases from which to attack enemy shipping and English towns on the south coast, although as members of the V111 Air Corps poured into the islands to be followed by men and aircraft from the 2nd Squadron Reconnaissance Group 123 (2 Staffel Aufklaerungsgruppe 123 or 'Battle Group Obernitz', after the commanding officer), one of the first things they had to do had little to do with fighting.

Jersey's potato and tomato harvest had to be transported and escorted to units of Luftflotte 3 and the Sixth Army in Europe. After this, the squadron continued their attacks against the British with venom. Initially those attacks had begun from Bretteville, in France, but between July 15 and 17, the remainder of the squadron arrived in Jersey, where they stayed until February 18, 1941. Initially they were successful, and according to German records, destroyed a barrage balloon a Spitfire, an E-boat as well as wreaking a certain amount of havoc

in coastal towns and cities. But the squadron's Dornier Do 17s were no match for RAF fighters and, although these were replaced by Junkers Ju 88s and Messerschmidt Bf 110s by the autumn, the RAF were giving any enemy aircraft they came across in the skies a battering.

If the RAF had not won control of the skies during the Battle of Britain, Jersey Airport might well have become a key factor in any German attempt to invade Britain, but we shall never know. What we do know, however, is that during the winter of late 1940, Aufklaerungsgruppe 123 was more interested in reconnaissance missions than fighting (planes were sent as far as Sheffield, Warrington and Liverpool, for aerial photographs) before the squadron was sent south, to the Mediterranean.

If anything, the skies above the Mediterranean proved more dangerous than those above the Channel, but the pilots and crew were not to know that as they prepared to leave.

Leslie Sinel's Occupation Diary of

Luftwaffe officers arriving in a Ju 52 at Jersey Airport

US Air Force officers with a captured Me 262 jet, one of which landed in Jersey en route to America

February 19, 1941, mentions their going: 'Practically the whole of the Air Force' was on the move, bound, he believed at the time, for Africa. All that remained was a skeleton staff who, in March, 1941, went on strike for danger pay and extra rationing (occasionally Allied planes would buzz the airport and drop a bomb or two on Nazi gun emplacements).

But the airport staff and Luftwaffe personnel were not to remain idle. They were ordered to increase the size of the airfield (which they did) at the same time as the amount of air traffic into Jersey decreased. This meant that there was more runway for fewer aircraft, highlighted by Leslie Sinel: 'Work at the airport has been accelerated and the airfield has been greatly enlarged' (March 31, 1942) and 'German planes now rarely seen and the airport is not being used very much' (July 4, 1943)

Perhaps it was just as well that few aeroplanes arrived — too many of them crashed as they tried to land — but, considering the few flights that were made during the latter stages of the war, the actual number of Luftwaffe personnel stationed here seems enormous.

In November 1944, for example, there were 1,533 officers and men stationed in the Island. Their main job was to man the 165 anti-aircraft guns dotted around Jersey, not to fly or maintain planes. As the war turned sharply in the Allies' favour, the airport began to see more action.

An early indication of this comes from Sinel's diary of April 30, 1944, when he writes about Allied planes flying over the Island and sounds of heavy bombing on the French coast.

He adds that a great deal of material had recently been flown out of the airport and that pylons had been erected in fields nearby and logs rolled on to the runway to dissuade Allied planes or gliders from landing. Mines had also been laid in the airfield — some of them to remain hidden for quite some time after the Liberation.

1940: German troops with a Junkers Ju 52

To a large extent, the pylons, logs and mines were no more than angry, futile gestures made by a beaten army. As the Allies continued to push their way deep into France, the occupying forces must have known that their stay in Jersey would soon be over. They were becoming increasingly isolated, and when German planes landed, gone was the easy pride of a conquering race. In its place were tired eyes and last-ditch efforts to protect the beleaguered garrisons of Lorient and La Rochelle.

To accommodate these troop planes, some of them taking soldiers into France, others bringing wounded men out, some of the pylons and logs had to be removed (for example, the runway had to be cleared when a Junkers Ju 188 stopped for a short while with letters and officers, bound for France, on December 23, 1944), but along with the mail and the men came the news that defeat was only a few, short battles away.

By May 1945, Hitler was dead and on May 5, resigned to their fate, the Germans cleared away all remaining obstacles on the main airport runways. On May 9, HMS Beagle rounded Noirmont and during the afternoon of the same day, the first Allied planes flew down from the skies — the planes being two American single-seaters whose pilots had popped in to 'have a look around'.

The war was over. On May 27, 1945, the first Jersey Airways aircraft arrived. By June, the first passenger flight had been scheduled, and on June 7, an RAF Dakota of 24 Squadron flew from Northolt to Jersey to pick up the King and Queen (who were in the island) to fly them back, via Guernsey, to the UK.

Two months later, a new age scudded to a stop on the grass runway in the form of a captured German Me 262 twin jet fighter. This was the first jet aircraft to land in the Island and a note in the cockpit gave a hint of the future with its warning 'Maximum speed not to exceed 700 mph'. Speeds in excess of 700mph, indeed! A new age might be dawning, but it was not going to arrive quietly, without fuss.

A Messerschmidt Bf 110, secretly photographed at the airport on its arrival in 1940

The Problem of Landing

IN February, 1940, a Fleet Air Arm Training Squadron was based in Europe for a short time using Fairey Swordfish and Fairey Albacore aircraft. They practised torpedo dropping in St Ouen's Bay and simulation deck landings at the airport at night.

Four months later and they had gone, to be replaced by the Luftwaffe, who used Jersey as a fighter air base until the Battle of Britain made them think again. Towards the end of the war it was used as a stopping-off point for Junkers troop carriers.

During the early days of the Occupation most of the aircraft flying into the airport were part of 2 Staffel: Aufklaerungsgruppe 123, known by the men as 'Battle Group Obernetz', after the first Luftwaffe commander in Jersey. By January, 1944, Staffel-Kapitan Obernitz had long gone, and anything flying into the airport was controlled by General-Major Gerlach, whose headquarters were at Bel Air, Saumarez Road, Guernsey. Indeed, Guernsey saw more aircraft land at their airport during the latter stages of the war than did Jersey, and this was probably just as well. For according to the late Leslie Sinel, writing in his war-time diary, 'The German Occupation of Jersey', Luftwaffe pilots mistrusted Jersey Airport — as the following shows.

November 1, 1940: German plane crashes at La Pulente while on a practice flight over St Ouen's Bay. The crew of five is killed.

November 3, 1940: German plane overturns and crashes at airport. There are two casualties.

November 5, 1940: In late afternoon a German scouting plane crashes in the grounds of La Rive, Rozel. There are two fatal casualties, the parachute of one airman failing to open, while the other airman is lost at sea.

November 7, 1940: One of the big German troop carriers crashes at the airport. There are over 20 casualties, the majority being pilots who had finished a course of training.

December 27, 1940: German plane crashes at the airport and is burnt out.

January 16, 1941: German troop carrying plane crashes into the sea off St Catherine. There are no survivors.

February 19, 1941: Much activity among the troops. Practically the whole of the Air Forces are leaving. And who can blame them.

A New Battle Commences

ON Thursday, January 15, 1948, during a debate in the States to decide whether £300,000 should be spent on extensions and improvements at Jersey Airport, Jurat Gallichan, president of the Harbours and Airport Committee, prophesied: 'I would not at all be surprised if, in the future, most of the traffic to and from the Island will be by air.' He was speaking at a time when the number of passengers flying to and from the Island had doubled (65,059 in 1946; 113,333 in 1947), and at a time when the airport was at the centre of not one controversy, but two.

First was the question of the size and cost of airport maintenance — £300,000 was an enormous sum to find, so soon after the war.

Second was a problem created by the new Socialist government in Britain under Attlee. A shrewd observer would have been able to guess the cause of this latest crisis when told that in 1947, 29,224 people flew on chartered flights and 84,109 flew with BEA.

So what had happened to Jersey Airways in the meantime? Immediately after the war, several faces familiar to islanders before the Occupation reappeared, including Charles Roche, who had been interned when the Germans discovered that he wasn't a Jerseyman.

Roche was appointed the airport's new controller, and G O Waters, OBE, who had been general manager of Jersey Airways in 1939, also returned to the Island, this time as managing director of Jersey Airways and her sister company, Guernsey Airways.

Although both Guernsey's and Jersey's airports had been enlarged during the Occupation, the two men had a lot to do. In Guernsey the airfield had deteriorated badly and was inadequately drained. In Jersey, any trees on or near the airbase had been cut down and used as fuel.

There were also land mines to find and

The airport terminal as the Germans left it

destroy. The main airport buildings remained intact, but the pre-war direction-finding and radio equipment, lovingly packed and shipped to England in 1940, had been destroyed in an air raid. It was mainly thanks to the RAF that suitable replacement equipment was found and installed quickly, at below market costs.

The airport was coming to life again and Jersey Airways were conscious that now was a good time to rub hands and count the money. As the only airline operating from the Channel Islands, their services would be invaluable for the next year or so as people and urgent supplies (one of their first flights was to collect 20,000 chicks to restock the Island's farms) were shuttled across the Channel.

On June 20, almost five years to the day since the last Jersey Airways flight had left the Island, a new, twice-daily service was announced offering flights from Jersey to Croydon via Guernsey.

At Croydon there was a coach waiting to take passengers on to Victoria Station and, although there were (as yet) no return tickets available, a single fare was just 85 shillings (£4.25).

The first flights from Jersey were made on two De Havilland Rapides (G-AGLE and G-AGLP) on loan from Railway Air Services and still in wartime camouflage colours.

With only one Southern Railways boat on the Channel Islands route, the need for more planes increased, and, during July, two newly completed Rapides (G-AGPH and G-AGPI) were made available to Jersey Airways, followed by further planes in August, October and November.

As the number of flights and planes increased, Jersey Airways switched from Croydon to Southampton, although office and Customs facilities at Southampton Airport were barely adequate. Despite such primitive conditions (it almost seemed as if the mainland authorities were operating out of a suitcase) during the first year of operations after the war, 1,430,069 lb of freight were carried by Jersey Airways and at times they were operating as many as 22 flights in or out of the Island, daily.

As Jersey Airways celebrated their phoenix-like rebirth and a new name, Channel Island Airways (although it remained Jersey Airways for many people in the Island), and as the 10,000th passenger on their new, 34-seater Bristol Wayfarer was given a meal, a free ticket to England and a perspex cigarette box (courtesy of the Bristol Aeroplane Company), Herbert Morrison's second reading of the Labour government's Aviation Bill was being debated in Parliament.

The fate of the railway companies had already been decided (they would be nationalised) and it was feared that all private air companies operating regular, scheduled flights in the UK would be nationalised, too. The Southern Railway Company and the Great Western Railway Company, who held shares in Jersey Airways and whose own fate had been sealed, were obviously unhappy about flagrant government intervention in a

business which served them and the islands well. But what kind of resistance could they offer against Whitehall?

Precious little, especially because the Socialists were determined that there would be no loopholes in their nationalisation programme. They wanted to do away with Jersey Airways monopoly of Britain's second busiest airport and to replace private air companies, registered in the UK, with British European Airways, British South American Airways and British Overseas Airways.

Airports would also be nationalised (the land they were on would be bought by the Government) to ensure that there was total state control of all flights over Britain.

It might have been easy to steamroller UK-based institutions into submission, but Prime Minister Clement Attlee had a problem. Could the British Government take over Channel Island Airways, which had been registered in Jersey, and could it place a compulsory purchase order on States-owned land?

After all, Jersey and Guernsey had their own government. The islanders didn't vote in British elections and, as more than one

States Member said on more than one occasion, 'the Channel Islands are an exception'.

Similarly, although Channel Island Airways were not the official island air company, they deserved different treatment. Airport and airline became a cause for stout, Channel Islands defence.

And so, for more than a year, there was a tussle between the States, Channel Island Airways and airport officials on the one hand, and the British Government on the other. At stake was the dubious right, claimed by the Government, to meddle in Channel Island affairs.

On February 26, 1946, one of the most important meetings between the two sides took place in London, where four representatives from the Home Office, three from the Ministry of Civil Aviation, five from Jersey and six from Guernsey met to discuss the matter. The Channel Island delegation were told that if they didn't agree to nationalisation, an Order in Council would be made, forcing the Civil Aviation Act on to the unwilling islands. The Government were determined that there would be no exceptions. However, the deputation managed to persuade government representatives that Channel Island land should stay in islanders' hands. The airports would not be nationalised, but they would have to be improved.

With the sad news that Channel Island Airways would go, if the results of this meeting were accepted, the delegation returned home and, for some time, kept quiet. But, as the year lengthened, it became obvious that some clear-cut statement about the future of aviation in Jersey was needed.

Senator Rumfitt, in particular, wanted to know what was planned, and he led the attack in the States on the British Government when the details of a letter, sent to the Lieutenant-Governor by the Home Secretary were made public. The letter explained that BEA would take over all regular Channel Island air services but the States could, if they so wished, operate a separate inter-island service under licence (BEA would be exempt from this licensing scheme).

On February 11, the letter was made public and on February 25, after Jurat Gallichan had introduced a variety of proposals in the States, to make the airport bigger and safer, Rumfitt launched into the

A Royal Air Force Dakota at Jersey Airport soon after the Liberation

A Channel Island Airways DH 84 at the airport

attack. He demanded that an Islands Air Company should be allowed to operate as an 'associate' of BEA, and a delegation was nominated to meet representatives of the British Government to press the claims of such a company.

Meanwhile, in Guernsey, the States had rejected the contents of the Civil Aviation Bill by 35 votes to nine, which was a much clearer margin than in Jersey, where it had been defeated by a marginal 23 votes to 20.

The Ministry agreed to review the situation (which they did), but on Thursday, March 20, 1947, a letter written on behalf of the British Government by the Minister of Civil Aviation made it perfectly clear that no matter how many times the situation was reviewed, it would not change.

Channel Island Airways must go — and by the end of the month the Civil Aviation Act (extension to the Channel Islands), 1947, became law. By April 1, the islands' local airline company, which had survived beach landings and the war, would be swallowed up by BEA, who had larger planes, more routes and already a great many ex-Jersey Airways staff working for them, including Commander Waters, who had joined BEA in August, 1946, realising that the days of small, autonomous airlines were over.

With hindsight, it was inevitable that the States had to bow to British Government pressure. It had been pointed out to a member of the Jersey delegation early on in the continuing debate that if the islands defied Parliament, no guarantee could be made that Channel Island Airways planes would be allowed to land at British airports. These airports were now owned, of course, by the British Government.

As BEA staff and planes arrived in Jersey it was natural, if unjustified, that they should be treated with initial hostility. It wasn't their fault that they had arrived on the wings of controversy— they were there because the Labour Government insisted that their laws be obeyed. However, one

Public roads still ran across the airfield and were controlled by traffic lights

faux pas which could have been avoided arrived at Jersey Airport soon after BEA took over. It was a German-built Junkers Ju 52 three-engined transport plane, fitted out by Shorts of Belfast as a civilian airliner and now described by BEA as their Jupiter class.

Its German origins did not go down too well with the Island, nor with the 'Evening Post' which took great delight in reminding everyone of its previous role in life. The last time that a Junkers plane of this type had been seen in Jersey was during the war, when it was used to ferry German soldiers to and from the Island. Could the British forget so soon?

August, 1948: The passenger waiting room

Making Way for Bigger Aircraft

WHEN Jurat Gallichan, president of the Harbours and Airport Committee addressed the States on February 25, 1947, he knew that the BEA v Channel Island Airways controversy would command centre stage. He was also clear, in his own mind, about the responsibilities he owed to his committee. Jersey Airport was unsatisfactory.

This had been known as long ago as July, 1946, when civil aviation experts had been called in to examine how well it had fared during the war. Although the Germans had made it bigger, they had allowed it to decline markedly during their final years in the Island. The pot holes on the runway were legendary and its landing equipment, although adequate, was not suitable for the modern airliner. The air navigation equipment, for example, was inadequate and it was suggested that radio and other navigation systems should be installed as soon as possible.

So Gallichan wanted better equipment. It made sense to buy as soon as possible, not only because it would aid incoming planes, but also because an up-to-date, all embracing communications system would help to prevent possible air (and sea) disasters. And, in their own methodical, highly efficient way, the Germans had set a precedent in monitoring the skies over the Channel. They had used Jersey Airport as a base from which to study aircraft movement even when no planes were using the airfield. So, one of Gallichan's priorities after the matter of who should use Jersey Airport had been sorted out, was all about equipment. Because safety was involved, the States might not grumble too long and too loud when he asked them for money for this. Unfortunately there was another matter which needed to be raised, and one which Gallichan knew would lead to controversy.

BEA, remember, were about to become

Larger aircraft would mean tarmac and a longer runway

the controlling influence over Channel Island skies. Their influence was far greater than Jersey Airways' had ever been and stretched the length and breadth of Great Britain.

Their planes also matched their size and status and it was inevitable that, sooner or later, they would want to introduce their larger aircraft on the Jersey route, in particular the chubby Douglas DC3 Dakotas, many of which were becoming available from surplus military stock. But BEA would not bring them to the Island unless the runway and the airport could cope with them. This wasn't just a matter of better equipment, it was also a matter of more runway which, in turn, would mean more land.

In early 1947 Gallichan formally asked that the Harbours and Airport Committee be given States permission to plan a larger, up-to date airport for the future. No money (as yet) was called for, and the States grudgingly agreed. With a few notable exceptions, most members wanted to keep the airport as their own, and there were always fears that it might be taken away from them by Attlee's aggressive Socialist government — but agreeing in principle and agreeing in fact were two different things.

An example of the kind of argument he was going to face when he approached the States with a huge bill for major extensions and improvements to the existing airport came in October 1947, when on behalf of the Piers, Harbour and Airport Committee (the 'Piers' was subsequently dropped), he asked the States for £1,326 2s 9d to buy Le Clos de Pipon, a field in St Ouen.

Everyone knew why the field was needed — as the 'Evening Post' headline explained at the time ('AIRPORT EXTENSION — More land to be acquired') but during the ensuing debate, the bitterness which many islanders felt towards this lengthening scar on the island's good, agricultural land was echoed on several occasions, not least by Jurat Collas, who had given his redoubtable opinion more than once.

'There is the question of agricultural land. And, as regards the size of machines, Their Majesties used the airport some time ago and if it was good enough for Their Majesties to land on, it should be good enough for their subjects.'

(March 19, 1947)

But Gallichan was adamant. And so on Thursday, January 15, 1948, he asked the House for £300,000 for telecommunications aids, a new hangar and for extensions to the main runway. Having delivered his message he stood back, and waited for the storm. It was not long in coming. So much money required, when the airport had cost a fraction of this amount to build? Why, it was scandalous! Throughout the afternoon 'a keen and spirited debate' took place within the House which tested Gallichan's knowledge (and nerve).

Thankfully, he knew his subject well, and was prepared for the onslaught. He had some staunch allies, too, including the superintendent radio engineer at the airport, J S Butchard, who explained that the airport had been living dangerously in recent years with as many as 218 aircraft movements a day and not enough sophisticated equipment to deal with them. Deputy Wilfred Krichefski (a man whose name will always crop up in any history of Jersey aviation) added his powers of debate in favour of the £300,000.

He, Butchard and Gallichan secured a moral victory that day in the sense that the States agreed to the proposals placed before them and even accepted Gallichan's plea for a longer runway (if the airport was to be of international standard it would need at least one runway with a length of 4,200 feet), however they could not agree to the cost.

As the 'Evening Post' explained at the time: 'During the course of the debate, Deputy Ed Le Quesne moved as an amendment that the sum be halved and, eventually, this was accepted by the president of the Harbours and Airports Committee and adopted with only one dissentient.'

Jurat Gallichan probably realised that £150,000 would not be sufficient, but once the building work had started, the States could hardly turn round and say 'no more', with the work half done.

A year later it was obvious that the job couldn't be finished unless the States did vote the Airport more money. It was at this stage that the new champion replaced the old, and Deputy Krichefski, in possibly his finest hour (and he had many) pleaded for an extra £105,000 so that Gallichan's dream would have a happy ending.

He mentioned, in his speech, that the States had already done remarkably well by buying St Peter's Barracks and 750 vergées of land from the War Office. The States paid £5,000 but it had a redevelopment value of £100,000.

He mentioned, in his speech, that the Island had to look to the future of its tourism industry — 130,000 passengers were already using the airport each year, nearly five times as many as in 1939.

He also mentioned that Jersey should be proud to own on of the best-equipped airports in the world and than an extra £105,000 would seem as nothing in the years ahead, when the Island would be reaping the benefits from money wisely spent, so many years before.

It would appear that no one had any suitable questions which would halt him in his tracks (this was one of the complaints which the 'Evening Post' made after the debate) but one question which Senator Rumfitt made, a question which was asked again and again in the 1950s, ran as follows: 'What would happen when the Dakotas weren't the largest planes around, a time when more land, more runway would be needed to accommodate planes twice or three times as large as the Douglas DC3? Where would the money and land be found to satisfy their huge appetites?'

In October, 1951, the Socialists were voted out of office and a new Conservative Government were voted in.

The UK was still recovering from the devastation of war (ration cards were still being issued, for example) while, in the Channel Islands, BEA were having to come to terms with a passenger trade which dipped alarmingly during the cold months of winter. Although their Jersey to London service was more popular than any other except London to Paris, the economics of running a year-round, regular schedule of flights was taking its toll. Full planes in summer couldn't compensate for empty planes in winter.

But BEA persevered and spread their net even wider. For example, you could fly direct to Birmingham, Manchester, Southampton, Northolt (which took over from Croydon in 1947), Guernsey, Alderney and France (although the direct Jersey to Paris route was abandoned in 1950 through lack of support). And, despite BEA's apparent monopoly of the air, other airlines were allowed to trade, under licence, at Jersey Airport.

It was a gateway to good business and firms like Jersey Airlines, whose first chartered flight was to St Brieuc in 1949, were keen to step inside.

They knew that BEA, whose fleet of 45 Rapides had dwindled to 19 by 1950 (the 19 were then known as the 'Islander class' within BEA) couldn't satisfy passenger demand at the height of the season and they weren't interested, anyway, in scheduled routes which fluctuated alarmingly during the year, which is one of the reasons why, in 1950, over 100 applications were received by BEA to operate charter flights out of Jersey. Meanwhile, in 1949, BEA had relinquished their early morning 'Pionair service' (the class name of the improved Dakotas) which brought in mail and newspapers to the Island, and sat back and suffered as their sternest critics (passengers and journalists) complained about the rising cost of air travel.

'Dear Sir, the British European Airways unilateral decision to raise the cost of air travel to the Channel Islands is typical of the despotism of monopolistic, state-subsidised octopi', ran the first sentence of a lengthy letter to the 'Evening Post' in March, 1956. It was a letter typical of many received by the paper over the years, but ignored some of the main reasons for price increases. For example, BEA were committed to serving an airport which could take 80,000 passengers in August, 1955, but a mere 7,000 in January the same year. And, although the new Conservative Government were keen to promote competition, they were not averse to a little direct intervention of their own. In the mid-1950s the Ministry of Transport and Civil Aviation brushed aside airlines' and travellers' complaints and demanded increased prices. Although representation was made to the Channel Island Air advisory Council (which features later in this chapter), in 1954 a weekend return from Jersey to London cost £11 11s (£11.55). A flat or small house could be rented, at the time, for as little as £1 a week.

BEA smiled grimly and carried on. New routes to Glasgow and Belfast were introduced, plans were made to buy shares in Jersey Airlines (whose managing director, Maldwyn Thomas, will feature again in the airport story), and the new Airspeed AS 57 Ambassador aircraft (the Elizabethan class and capable of carrying 49 passengers at 235 mph) flew into Jersey.

The airport was open for 21 hours at the height of the season and Jersey seemed a million miles away from those early days of 1917, when Gertrude Bacon had tentatively hinted that air transport had quite a future.

On December 12, 1947, the Channel Island Air Advisory Council met for the first time. With delegates from Jersey, Guernsey and the British Ministry of Civil Aviation, its main job was to look after the future of Channel Island aviation and to serve as a link between Island and UK Government policy. The year before, in July 1946, a 17-year-old girl had learned to fly in a De

1947: The airport from the air

Havilland DH 82a Tiger Moth biplane. She was the first successful pupil in Jersey's newly formed flying club. In September that same year, the RAF Association of Jersey pioneered a Battle of Britain air display as a fleet of RAF Mosquitoes took to the skies on September 8.

Three unrelated events, but they show to what extent Jersey now accepted the importance of air travel and, by implication, the airport — one which had been improved, dramatically, in the late 1940s and was to change again over the next few years.

'I WAS talking to a friend, referring to my fear that a hot summer sun would affect tarmac runways, he tells me that during 1943 and 1944 on the occasion of very hot weather, 90 per cent of the aerodromes in Britain with tarmac runways were out of commission.'

(Under the Clock, 'Evening Post', April 14, 1951)

The first call for a tarmac runway was made on April 3, 1951, when the Harbours and Airport Committee asked the States for £47,000 to build one. The money and the tarmac were needed and, despite reservations about which was best, tar or concrete, the States agreed that a 4,200 foot tar macadam runway be built. In May, 1952, it was finished. Five years later it had been extended to 4,550 feet (hardcore for the runway came from a dump in St Peter's Valley, building material which had been excavated by the Germans during their digging out of the Underground Hospital). By this time most islanders appreciated the airport's value to the community.

When the 1955 air/sea figures were published in January, 1956, it was revealed that while 354,416 passengers arrived or left Jersey by sea, 383,527 used the airport. For the first time air travel was more popular than sea as a means of getting to and from the Island.

The new runway could take BEA Elizabethan aircraft or Vickers Viscount 701s, which BEA were buying, and Jersey's unique position as a holiday island, close to France but English-speaking, with duty-free goods and carefully graded hotels and guest houses, made it a tourist paradise, or so most holidaymakers believed. By 1957 the 'Evening Post' recognised the airport's value to the community.

'It is fairly certain that were there no air services, hardly one quarter of the present number of tourists would make the journey by sea. Thus it is that Jersey's prosperity and even her standard of living become even more dependent upon the maintenance of efficient airline services to and from all parts of Britain and upon the growth of those services as public demand warrants.'
(Leading article, February 20, 1957)

It was good to have the local newspaper on the airport's side, particularly as Senator Krichefski would soon be seeking money and pushing through legislation to guarantee its future.

1953: During the air display at Jersey Airport Senators Krichefski and Le Quesne tried out another innovation, and one which had nothing to do with the question of noisy jet engines — a helicopter

The Jet Age

THE runway needed working on. Load-bearing tests and classification of airfield payments in 1958, conducted by independent experts from the Air Ministry, had resulted in a report suggesting that one and a half inches of asphalt top-coat needed to be added to all airfield pavements plus a thick tar macadam base to the 500,000 sq ft Jersey apron.

According to Krichefski, the Harbours and Airport Committee would have submitted the relevant Bill in 1958 if the States engineer had not suggested that they wait a few years. After all, he pointed out, there was still a great deal of wear left in the existing tarmac. Now, however, the time had come.

On July 8, 1964, Krichefski, president of the Harbours and Airport Committee, asked the States for an extra £20,000 to extend the eastern end of the main runway. 'The extension has no sinister background at all,' Krichefski explained, 'some people have said it is rumoured that the runway has to be extended to cater for jet aircraft. This is not true'.

The money was granted and the extension allowed. The president's explanation was true — there were no plans for jets to be invited to Jersey —BUT the main companies operating to the Channel Islands, including BEA, British United and Derby Aviation (the forerunner of British Midland), knew that as the main runway increased in size, it COULD cope with jet aircraft, particularly as a massive £550,000 had been promised for the airport's runway, taxiway and apron just three months before.

Between 1960 and 1964, the Senator told the States, 2,420,000 tons of 'all-up' weight of aircraft had used the airport. Since 1952, when builders replaced grass with tar, the runway had been extended from 4,250 feet to its current length of 5,100 feet (it was 150 feet wide) at very little cost. By comparison, Krichefski pointed out, Guernsey's 4,800 foot runway, only 125 feet wide, had cost £120,500 — £40,000 more than Jersey's own, longer and wider runway.

On April 4, 1964, the money was approved, despite Deputy Maldwyn Thomas's complaints that the States should invite more experts to the Island to assess the current state of the runway and taxiways. States Members, it would appear, were confident in Krichefski's knowledge of his subject. After all, he had been closely involved with the airport for 17 years now and they trusted his judgement. He also told them that the improvements would, ultimately, cost the Island nothing. If landing dues were increased, by 1979 the States loan would be   paid back, with interest.

It was another personal triumph for Krichefski, and he must have been delighted by the way the States received his proposals. They, at least, were prepared to listen to him, which couldn't be said when he had to meet parishioners from St Peter.

For years he had enjoyed a love/hate relationship with the parish. He loved their airport, but a great deal of the people who lived close by hated the sound of aircraft landing or taking off from the extended runway. They also hated the amount of land that the airport continued to need, and were never loath to say so.

Money was rarely the problem, but land was. As the airport grew and grew and as legislation controlling civil aviation tightened up, any house or outbuildings on the main runway approach became potential hazards and had to be removed. In 1954, for example, when a St Peter's man had had his plans to build five bungalows on a site to the eastern end of the runway turned down, complaints to the 'Evening Post' continued long and furious and, in 1957, there was a full scale, noisy debate when senior officials from the Ministry of Transport and Civil Aviation Authority recommended that 11 houses ought to be demolished.

Safety, the States decided, was more important than land. The houses would have to go, as would La Commune Farm, plus a great deal of agricultural land which was needed to make space for the B36, a new road from the airport to Beaumont, called not surprisingly, The Airport Approach Road.

With the help of the Constable of St Peter and Deputy W J Simon, the reluctant landowners were persuaded to sell, but they, and other parishioners living nearby, were concerned about their future. The land around the airport had been placed in different zones and those islanders with property in the White Zone (a dumbbell shaped safety area of land surrounding the approach to the runway) were worried that any requests to build or alter existing buildings on their property would be vetoed. The Natural Beauties Committee (later to be called the Island Development Committee, or Planning), would refuse development permission whenever the Airport said 'no'. Even a desire for an outside toilet could cause controversy, as Senator Krichefski discovered in September, 1959.

He was vilified at St Peter's Parish Hall one late September evening by parishioners who were fed-up with owning land which was of no use to them. One landowner, for example, wanted to know why the airport had objected when he submitted planning permission to build an outside toilet. Another complained bitterly because he had been refused permission to erect a vegetable frame on his white-zoned land.

In reply, Krichefski pointed out the necessity for keeping the area as flat and as uncluttered as possible. It was not his desire to say 'no', but safety precautions with regard to all incoming air traffic took priority. He repeated what he had said at a similar angry St Peter meeting two years earlier.

'A great deal of work has been done at the airport since the Occupation, and, probably, had the money and land been available then, it would have been better to re-site the airport altogether. . .'
(July 15, 1957)
He may not have convinced his audi-

The waiting room at Jersey Airport during the 1950s

ence but, in an island dependent on tourism, he could rely on other States Members, hoteliers and islanders involved in the tourist industry to do the talking for him. Senator Rumfitt pointed out that the airport meant more than £5 million a year to the community.

Despite complaints about restrictions on land, the usual complaints about rising air fares, and an odd complaint against BEA in 1959 that it was wrong of them to sell bus tickets from the airport to St Helier on their flights, because of the effect that this would have on taxi drivers, the years before 1956 and 1964 were remarkably good ones.

Air arrivals continued to increase, and although 1963 saw a slump in the overall number of tourists to the Island when compared with the previous year, airport arrivals were up (from 414,536 in 1962 to 417,573 in 1963) while sea arrivals were markedly down (146,610 to 127,740).

Holidaymakers were flying in from as far away as Sweden (350 arrived on charter flights in 1960) and Channel Islanders were also taking the opportunity of using the airport to jet away to the sun.

'Eleven day, 15-day and 19-day holidays to Morocco and southern Spain, including four days on the Durban Castle from Gibraltar — 56 guineas (£58.80)'

'Majorca. October 9 and November 6, by air from Jersey Airport by Viscount airliner. Selection of five hotels. Optional sightseeing and evening entertainment. Fifteen day holidays from 46 guineas (£48.30).'

Gaytour Holidays, 1960)

If holidaymakers were delayed (there was some notoriously bad weather in the early 1960s) they could find some comfort in a departure lounge which had been built as part of a three-stage rebuilding programme, begun in 1956, at a cost of £125,000. When the work was finished, 1,100 passengers could wait in comfort until the fog had cleared, compared with a miserly 410 a year before.

Meanwhile, a few bonuses, not of particular importance to the general public, perhaps, but important to the Island and to air

Maldwyn Thomas and hostess Diana Berkelman stand in front of a Jersey Airlines Dakota (below) and (above) hostesses in the airline's new uniforms in July, 1956

The Elizabethan was the largest passenger aircraft of its time to operate in Jersey

traffic control added to Jersey's high standing in the world of aviation and commerce. Freight services improved. Prime Jersey cattle were flown out of the Island to new herds in early every developed country in the world. Growers sent their produce direct to market in huge quantities. On March 19, 1963, for example, 11,053 kilos of flowers left for mainland distribution.

In February, 1959, the airport's love affair with modern technology reached a new height with the installation of the £54,000 Marconi S264 long-range surveillance unit, which meant that it could control all flights within a radius of 45 miles, to a ceiling of 20,000 feet. In November the same year, the Decca Navigator system was tested on incoming flights (at no cost), making the Jersey route one of the safest in the world. To this day the airport has one of the most enviable safety records, despite three crashes (the last one as far back as 1980) which affected the Island community for many months afterwards.

As 1963 became 1964, nobody knew what the future held in store, but at least 11 parishes were filled with residents who appreciated what the airport meant to them. Some measure of the increasing prosperity to Jersey brought about by tourism (and indirectly by the role that the airport played in this industry) can be gauged when the £550,000 requested by Krichefski for airport improvements is set against the £7 million allocated on Budget Day, 1964. This was the amount the States agreed for ALL island expenditure. No one could ignore the Airport's importance, although accompanying prosperity came larger, faster and noisier aircraft. In November, 1964, it was rumoured that British United Airways planned to fly the new BAC 1-11 (the bus-stop jet) on scheduled passenger services between Jersey and Gatwick in 1965 if the Air Transport Licensing Board gave permission.

British United had ten of these new, 540 mph BAC 1-11s on order and it would have pleased management at the time that Jersey's main runway was being lengthened and strengthened. The main runway could just about take a jet aircraft, even if some of the people living in St Peter couldn't.

'HERE, inside the house and almost a mile from the airport, the actual, physical shock vibrated like a road drill against the body. I felt physically sick and mentally desperate for several minutes and a full hour later when our lawn mower started up, I was appaled to hear myself scream "Stop it! Stop it! I can't bear any more noise!".'
(Veronica Platt, letter to the 'Evening Post', August 24, 1965).

— Jets were landing at Jersey Airport.

Despite BUA's claim that 'BAC 1-11 bus-stop jets were quieter than most jet planes because of the rear mounting of their engines, and that their steep and rapid climb from the airport would eliminate any noise problems', St Peter residents were not convinced.

The old enemy, Senator Krichefski, was branded as a traitor and, after the first 1-11 arrived at the airport, bringing travel agents to the Island on August 17, 1965 (the Bournemouth to Jersey trip took 21 minutes), there were meetings at St Peter's Parish Hall most weeks as residents took up the cause of noise pollution. And nearly every day a letter arrived on the editor's desk at the 'Evening Post' either for or against jet aircraft and Krichefski.

The Senator's statement that 'jets are inevitable' in September, 1965, only fuelled the controversy, and in November the guardians of St Peter, the Airport Zone Residents Association for the Prevention of Aircraft Disturbance, was replaced by SPAD (the Society for the Prevention of Aircraft Disturbance) in a bid to keep the parish a jet-free zone.

It was not a time for the faint-hearted as Captain G Thomas, BUA operations supervisor (Gatwick), discovered when he was invited to St Peter to talk to the residents.

'A rowdy and sometimes rude meeting was held at St Peter's Parish Hall last night to discuss the noise problem of the aircraft, the BAC 1-11. A guest speaker (Captain Thomas) was at one time ordered to sit down by the audience of anxious residents who filled the hall to standing.'

(October 30, 1965)

In the middle of this heated debate, Channel Airways let it be known that they were thinking of using jet aircraft on the Jersey run. BEA were also 'assessing the situation' and everyone realised that once the door had been opened for one airline to use them, others would follow.

However, no one seemed too sure about the hazards of these jet-propelled monsters.

'Is the paraffin fall-out from a jet aircraft greater or less than that of a jet propeller, and is it not a fact that paraffin is damaging to growing crops, clothing and drinking water?'

This was just one of the many questions fired at Krichefski during his visits to parishioners living near to the airport and the Senator was never afraid to try to answer them ('there is no proof that paraffin fall-out is greater from a jet, or more harmful').

Unfortunately, the question of noise could not be explained away so easily.

On December 14, 1965, the matter was debated in the States when noise levels as high as 118 decibels, registered near St Peter's Church, were mentioned. But, as usual, Krichefski had done his homework.

He had consulted many experts, including the Deputy Secretary at the Ministry of Aviation and the Southern Divisional Controller with responsibility for technical services at Gatwick, over noise levels.

'Yes,' he admitted, 'the maximum level at Gatwick was 110 decibels and yes, it might be that, occasionally, an extra high reading near Jersey Airport could be recorded. BUT it wasn't an "even noise". In some areas near London's main airport you could obtain readings much higher than 118 decibels, in other areas, much less. The same would be true of Jersey.'

Over and over again he was heard to say: 'I didn't put up Jersey Airport in St

St Peter's parishioners protest against plans for a larger airport

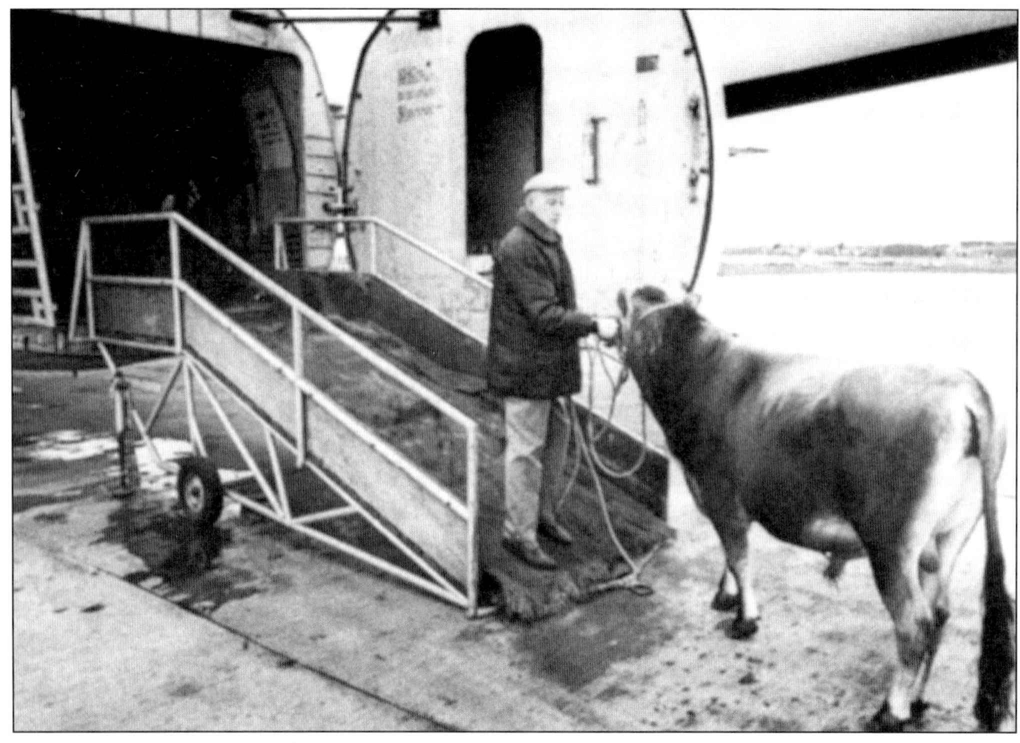

Jersey cattle have been flown throughout the world from Jersey Airport

Peter in 1937. Jersey Airport IS at St Peter and there is nothing we can do to alter that.'

It was a persuasive, decisive speech and it only just worked. By 25 votes to 24 the States agreed that four BUA jet flights would be allowed daily until March, 1967. SPAD delegates licked their wounds and returned home to consider the next move in their anti-jet campaign.

As far as they were concerned, the first skirmish had gone the way of their bulky enemy, but the war would continue until an honourable decision (in their favour) had been reached. Experts were invited to the Island to 'prove' that jet aircraft would do immeasurable damage to such a small community. SPAD took on a new chairman (Advocate Perrier), and an Airport Consultative Committee (whose main complaint was that it was never consulted) was formed with Krichefski's help.

Islanders living near the airport hoped that they would be able to influence airport policy through the committee; Krichefski wanted it to be a means of explaining that policy, to placate any parishioners without allowing them too much say in any decisions taken by the Harbours and Airport Committee.

Despite the fuss, jets would become a part of Jersey's way of life and it was inevitable that once one jet had landed, more would follow. On March 22, 1966, the States agreed by 38 votes to 12 that current restrictions on jet aircraft would be removed by March the following year.

In July, 1966, the Hawker Siddeley Trident jet (owned by BEA) arrived in Jersey for tests. In June, 1967, Channel Airways' 90-seater BAC 1-11 series 400 arrived from Southend and, in July, 1968, their 139-seater Trident, operating out of Stansted, touched down on the runway.

In November, 1969, the 'Evening Post' reported that 'larger and more powerful BAC 1-11 jets have been flown in and out of Jersey but the increase in noise seems to have gone largely unnoticed by many islanders . . . and, to an outsider, it might have seemed that the airport was a law to itself.

Tragedy

Sous l'Eglise: On Wednesday, October 1, 1982, 50-year-old Norman Harvey was returning to Jersey in his private, twin-engined Cessna Citation. The plane approached the airport low and off-centre through the evening gloom and, as Vernon Balleine, of La Chasse, St Peter, said afterwards: 'I heard the aircraft pass directly overhead and, a fraction of a second later, saw the sky light up.'

Mr Balleine immediately made his way to the burning aircraft which had crashed into an old granite house, Sous l'Eglise, near St Peter's Church skidded into outbuildings and exploded. There was nothing he could do but, mercifully, no one except the pilot had been killed or harmed.

This was the only air fatality during the 1980s and 1990s, and was due to pilot error.

Oak Walk: On Wednesday, April 14, 1965, a British United (CI) Airways Dakota callsign G-ANTB, piloted by Captain Peter Self, a Guernseyman, took off from Orly Airport at 5.32 pm and made his way through fog and cloud to Jersey. Shortly after 7 pm, in trying to land, the plane came in too low through the fog, struck a landing pole at Oak Walk, St Peter, cartwheeled into a ploughed field and burst into flame.

All 23 passengers and four crew should have perished but, remarkably, 22-year-old air stewardess Dominique Sillière was found by rescuers lying on her back near the remains of the tailplane. She was dragged clear and apart from two broken legs and no real appreciation of what had happened ('I shouted to the passengers', she recalled afterwards, 'but remember only silence') she survived, unscathed.

By coincidence, the landing pole which the plane had struck on its way down had only recently been given a new number. Its previous one? Thirteen!

St. Peter's Farm: 'It is with deep regret that the company have to announce that one of their aircraft, G-ACZN, known as the St Catherine's Bay, met with a fatal accident at 10.52 this morning after leaving Jersey Airport for Southampton.'

(Official statement, Jersey Airways, Friday, November 4, 1938)

Twenty-seven years before the tragedy at Oak Walk, another plane had crashed, carrying passengers. Thirteen people died, plus Mr Edmund Le Cornu, foreman at Mr Lauren's Farm, St Peter, who was hit by a fragment of the disintegrating plane as he was working in the field. And while we know why the Dakota crashed in 1965 (because of poor visibility) the reason for this 1938 disaster remains a mystery. The plane made a normal take-off, circled in the direction of Red Houses and then, without warning, crashed into a field about 600 yards to the right of the airport, bounced across a small country lane and burst into flame.

Although one of the people who witnessed the crash believed that the pilot, Mr A G M Carey, was trying to land in the field, and more than one onlooker commented on the way that the plane tipped suddenly forward, propellers thrashing the air madly, why it fell to earth remains a mystery.

The Cruel Sea: On July 31, 1936, The Cloud of Iona, a twin-engined passenger plane, piloted by Captain William Halmshaw, took off from Guernsey just after 7 pm, bound for Jersey. The aircraft was next seen in pieces, wedged between Les Pipettes Rocks at the Minquiers, 13 days later.

The bodies of the two crew and eight passengers were eventually found in the Bay of St Malo, but it was obvious that the plane had not crashed — the ten people had drowned — which suggests that the plane (an amphibian) had landed at sea. At the time it was rumoured that she had come down near Corbière, close to the mailboat, which was steaming towards St Aubin's Bay, but only the sea will ever know.

This was far, far from the truth. SPAD had been busy behind the scenes and they and opponents of any unjustified airport extensions were to gain a kind of victory just three years later.

In the autumn of 1968 a Board of Trade report on aircraft noise in Jersey told islanders what they already knew — 'there is no easy solution to noise' — and, more helpfully, suggested that banks of earth should be placed around the runway to act as shields, to stifle sound. Their report also advised that a noise monitoring system be established, building development be more carefully controlled and decibel levels be carefully monitored.

The worry of keeping the area around the Airport clear of houses had not escaped the Harbours and Airport Committee, who had already agreed with the Island Development Committee that a conical shape around the airport (4,300 feet at its widest point, tapering to 2,290 feet near St Peter's Valley) should become a 'sterile area' for which no planning permission would be given.

The reasons for this sterile area were twofold. Although it was smaller than similar, undeveloped areas of land around UK airports, it gave pilots more clear sky as they took off or landed.

And it also gave Krichefski room to manoeuvre.

Always looking to the future, he realised that the main runway might have to be lengthened perhaps several times in future years (BEA had been pushing for another runway extension in November, 1966) and he wanted to ensure that such proposals would be accepted by the States without too many hiccoughs — which is another reason why the Harbours and Airport Committee were buying land (as it became available) on all sides of the airport's boundaries.

Such far-sightedness had its advantages.

After considering the Board of Trade report in early 1969 and after consultation with other air safety officials, the Harbours and Airport Committee lodged au greffe (to be debated in the States) a series of proposals which linked air safety with land usage and noise levels.

After a series of discussions inside and outside the States, it was decided:

a) That there would be three planning control zones, Zone 1 to be most tightly controlled, with only essential buildings allowed; Zone 2 which would be carefully monitored and on which there would be no major residential development; Zone 3 to be an area where development would be allowed, although it would be under constant review.

(An amendment by St Peter's Deputy Maldwyn Thomas, agreed by the States, meant that these zones would be considered again three years later. They were — and in October, 1974, Noise Zone 2 was redefined so that another 40 houses could be built in St Peter's Village.)

b) Planes would now have to fly 1,000 feet before cutting thrust (it had been 500 feet before, which had been too low, too dangerous).

c) The suggestion that the noise levels should be confined, by law, to 102 PNDB (decibels) by night, 108 PNDB by day, made by the Board of Trade, was debated but Krichefski reassured the House that the noise levels were already lower than this and reminded States Members that no jets were allowed to fly after 10.30 pm, that there was a general embargo on ALL aircraft between midnight and 7 am — except, of course, in extenuating circumstances.

If SPAD members weren't over-impressed by these decisions and by the speed with which jets were accepted into the Island, everyone in Jersey was conscious of the need for greater safety, and there were worse things to contemplate than excess noise. This had been brought tragically home to the community on Wednesday, April 14, 1965, when a BUA Dakota (GANTB) crashed in dense fog in a field near the airport.

A Change of Direction

'IF we worried about Jersey, the next on the list would be the ships on the Atlantic and every chicken that's frightened into laying square eggs.'
(Air France, December 13, 1967.

Islanders were already writing to the company complaining about the damage that might be caused to their homes if Concorde regularly flew overhead.) This chapter is not about noise, although the thought of Concorde sonic-booming its way through Jersey skies must have horrified some islanders, already fed up by the noise of jet aircraft taking off at regular intervals from the airport. Instead it concerns the odd events of May 1972, when a new president of Harbours and Airport came suddenly to power and a newly-elected committee began to look west, instead of east.

As Deputy John Ellis eased himself into the large chair recently vacated by Wilfred Krichefski he must have realised, all too soon, that it was not as comfortable as it first appeared. The discomfort was initially felt in February, 1970, when Ellis wanted Senator Krichefski to remain on the Channel Islands Air Advisory Council, despite the fact that he was no longer president of Harbours and Airport.

Some States Members disagreed and said so, quite forcibly. The new president must have been surprised at their hostility but his reasoned argument that continuity would be good for the council made sense and, after a tussle, Krichefski stayed.

Soon afterwards, Deputy Ellis was asked to consider the implications of the Edwards Report (which was to have serious implications for BEA) and, as was wading through this government-sponsored assessment of current and future aviation policy, he was called upon to justify the siting of an instrument landing system in a field next to Rue des Landes, which could have meant closing the road altogether.

In May 1970, he outlined to the States the importance of this type of equipment and, despite more grumbles about the tak-

ing of land and noisy jets, he survived until the New Year without too many problems.

In January 1971, a circular from the Department of Trade and Industry gave him new cause for concern and the possibility of a few more sleepless nights.

According to the circular, the safety area at the end of the main runway was 200 feet when it should have been 300 feet longer. So, as president of the Harbours and Airport Committee, Deputy Ellis was asked to present a full report to the States.

He did so and a 600-page document was presented in December, 1971, for discussion. It contained some revolutionary ideas which would have altered the look of Jersey Airport forever. Among some of the more contentious suggestions were the following:

a) An extra 300 feet would have to be added to the runway if the Island were to meet the UK requirements.

b) If another 300 feet was added . . . then why not add 1,000 feet, instead, for the difference in cost between 300 feet and 1,000 feet was not excessive.

c) The cost of such an extension plus other safety requirements would be about £1,250,000.

It had taken a long time to compile the report (one of the criticisms levelled against it) but at first glance the recommendations it contained, though radical, seemed persuasive and necessary.

If adopted, they would make the airport a safer place and, coincidentally, would enable BEA to fly their Super BAC 1-11s in, to replace their Viscounts.

But they created uproar in St Peter and her bordering parishes. At a specially convened meeting at St Peter's Parish Hall, for example, a resolution was passed urging that an independent enquiry be set up immediately on the grounds that: 'We do not want this island to become an aircraft carrier . . . the aircraft that fly to Jersey should be made to fit the size of our little island . . .'
(Mr A Flim-Hurtle of St Peter's Valley, January, 1972)

In St Mary the reaction was the same, although the parishioners could offer few solutions to the airport problem. However, one idea that 'the runway be placed at a different angle' was debated at a parish hall meeting in February during which a parishioner added 'it seems that everyone wants a runway so long as it does not affect them' — a true comment, but of little practical use to either the Harbours and Airport Committee or Deputy Ellis, who found that his lengthy report had stirred up a hornet's nest of opposition. This time that opposition seemed to be Island-wide . . . it wasn't just one parish in 12 which deplored radical change.

On May 24, 1972, the hammer fell. It was wielded by Senator John Le Marquand, who told the States that he had studied the report for several months, had consulted the Civil Aviation Authority and the British Aircraft Corporation and had, reluctantly, come to the conclusion that the report was full of holes, not least because it was trying to abide by rules formulated by the British Civil Aviation Authority and not the Harbours and Airport Committee, who were the real rulers of Jersey's air space and airport. Was a new, 1,000 foot runway extension really necessary? Were the taxiways in the wrong place? How realistic were these plans, which had taken so long to draw up and which would require a great deal of Jersey land and Jersey money to turn blueprint into fact?

Deputy Ellis found it almost impossible to answer Le Marquand's questions and, after a vote of no confidence was called (which was passed by 32 votes to 17), he resigned. He was replaced as president of Harbours and Airport by Deputy Bill Morvan, a new committee was formed (it included John Le Marquand), and the States were told that the many recommendations in the report would be dropped, henceforth.

St Peter breathed a sigh of relief (which was half-choked when they learnt that Rue des Landes would be taken away from them and closed, in the interests of air safety, later in November). BUT it was partly their pressure which had led one deputy to be replaced by another as president of a committee that had, for so long, been associated with controversy.

'What,' they wanted to know, 'would Bill Morvan do now?' After lodging a proposition that £210,000 be allocated for a new radar system, Deputy Bill Morvan and the new members of the Harbours and Airport Committee considered again the whole question of the Airport . . . and waited.

There was going to be no mad rush to take away more land from St Peter to bring larger jets into the island, to aggravate the parishioners at excessive cost. Even when the pilots' union, BALPA, attacked Jersey Airport, saying it was 'studded with alarming problems' in 1973, adding that they were 'disappointed' that recommendations made by the Department of Trade and Industry hadn't been implemented, Bill Morvan stood firm. He admitted that work needed to be done, told the press that '17 million people have taken off and landed there safely since 1947 . . .' and waited. He wasn't going to be rushed and wanted to know what decisions would be made at the International Aviation Conference held in Montreal, in November (one of the topics under discussion was runway lengths) before deciding on the airport's future.

Four months later, on March 26, 1974, the Harbours and Airport Committee acted.

A new, £1 million safety plan was presented to the States which was designed to make the runway longer and safer without causing the unnecessary aggravation of previous plans. The minimum amount of land would be taken away from islanders and, this time, the whole concept of the scheme was truly revolutionary . . . instead of taking another chunk of St Peter on the east of the main runway, the plan depended on the feasibility of taking land to the west. One of the main propositions was that there would be a new, 450 foot embankment on the St Ouen's side of the airport, where the terrain dropped sharply to the bay.

After a huge amount of back-filling and underpinning, new land would be created

Deputy Ellis

where previously there had been just air. Morvan and his engineers believed it would be a costly business, but it would work. The extra extension would mean that if the runway threshold was taken back 100 feet and if another 100 feet was added at the eastern end, after development the total runway length would be about 6,000 feet. There would also be the required 500 foot safety area at the western end. At the moment, the runway conformed to international safety standards but, when the work had been finished, it should conform to more stringent UK standards as well.

Quietly, on a standing vote, the States approved the scheme. Despite problems in the west, where 'construction poses a serious problem because of the land falling away in a steep slope' (a progress report in June, 1975), and despite the problems of a reduced runway as work continued, by early 1976 much of the work had been done, the Instrument Landing System had been resited, secondary radar (enabling air traffic controllers to identify planes and to read their heights) had been installed, and 300 feet of concrete, 14 inches deep, was being laid in the east.

And, with SPAD acknowledging that the new Harbours and Airport Committee had their interests well to the fore (in February, 1976, one of their members, Mr J V Schilling, explained that 'We feel there is much less need for a pressure group like ours, now that we have a sympathetic president. . .') it would appear that the Harbours and Airport Committee and the new runway had won over admirers from a previously unhappy parish which had seen everything from Bristol Wayfarers to the BAC 1-11 (500 series) land in their midst. Meanwhile, what of the old airport champion, Senator Wilfred Krichefski, who must have studied Bill Morvan's plans to head east with keen delight?

On December 12, 1974, while driving to the airport, he suffered a massive heart attack and was subsequently involved in a head-on collision with an oncoming lorry. Death was instantaneous. For nearly all of his adult life he had loved, fought for and cherished Jersey Airport and it seems fitting that he was on his way to this 'second home' when he died. He never lived to see the new runway replace the old or to appreciate how Bill Morvan's plans to extend the safety area eventually paid off.

For 'pay off' they did, especially if life means a great deal more than noise and money.

Deputy Morvan, who decided to build west, not east

The airport main building increased in size in the 1960s and 1970s

Concrete foundations are laid as the airport grows

On Saturday, August 20, 1977, a Viscount aircraft with 77 people on board came to rest only a few feet near the end of the 500 foot 'overshoot' to the west of the runway. So near to disaster had the plane travelled that its nose was in the 'arrester pit' — about a yard away from toppling into St Ouen's Bay.

With ponderous tones the 'Jersey Evening Post' explained:

"Had the accident happened a year before (ie when work on the overshoot was continuing) the plane would undoubtedly have crashed down the steep slope beyond, with heavy loss of life.'

That the plane didn't crash down the slope was mainly thanks to Bill Morvan's perspicacity. He cared about his airport and its excellent history of air safety . . . Senator Krichefski would have approved.

'Terrazo Flooring and White Ceilings'

'IF the committee of 1937 which opened the airport were present to see a proposal to provide a passenger pier, they would have believed that an H G Wells' fantasy had come true . . .'

So said Senator Bill Morvan, president of the Harbours and Airport Committee in June, 1976, during a debate which was to lead to a new passenger pier being built and opened on March 19, 1978. On that rainy, March day, the first passengers (127 Danes who arrived on a Boeing 737) were all given a glass of champagne to celebrate. And grander plans were in the air.

In March, 1979, it was rumoured that a new departure hall would be built above the existing departure area (which would then become the main arrivals and baggage handling area).

The rumours were true, but the plans were scrapped because, as Senator Bill Morvan explained, 'they are too expensive in the current economic climate'. Not to be deterred, however, the Harbours and Airport Committee went back to the drawing board and thought again. And what had those 1979 estimates been, for a second storey passengers' area at the airport? . . . £1.9 million.

In 1937, remember, the initial estimate to build a complete Jersey Airport had been £19,000. How times had changed . . .

Once it had been announced that the airport would be built at St Peter and within months of the first plane touching down on the runway (an unscheduled landing by a Jersey Airways' 14-seater, on November 7, 1936), it was inevitable that it would grow in popularity, that more planes would arrive, more passengers would disembark and that more passenger accommodation would be needed to house them.

In 1956 Senator Krichefski had anticipated a busier airport when he asked the States for £125,000 so that, among other improvements, a new passenger hall to accommodate 1,000 could be built. The money was approved. Within ten years, however, it had become obvious to any passing traveller that the facilities at the airport were, once again, inadequate, which is why a new departures lounge, costing £816,000 with 'floor-to-ceiling glazing contained in aluminium alloy fitments, with terrazo flooring and white ceilings (to give the illusion of great height)' was built in 1965. As well as being ultra-modern, it incorporated a room where mothers could take their young children, to change them and wash them, while waiting for a flight.

But passengers wanted more than somewhere to sit, or space to change their children's nappies and, almost immediately after the new departures hall had been built, plans for another extension — upwards — were revealed.

In 1968 the States were asked for £180,000 so that two new wings could be built above the existing administration block. Approval was given and on Tuesday, August 4, 1970, a new, 175-seater restaurant and bar were opened to the public. On the same day a new air traffic control radar room and control tower (to replace an old, weather-beaten shed, at the top of the building) and a badly-needed meteorological station (at a lower level) were officially used for the first time.

The look of the airport was changing and, with a lift connecting the ground floor (where a shop had been opened since 1969) plus other, smaller changes, passengers were at last getting the kind of priority that the main runway had been getting for so many years.

'It was about time that passenger comfort and the future of passenger comfort were considered,' Brian Mellor, the airport commandant, explained to the press that August day. 'By 1980,' he continued, 'two million passengers a year will be using the airport . . . already 72,000 aircraft movements and 10,000 tons of freight were being handled each year. . .' which went some way to justifying the recent costs and the 215 States-employed workers who ran the airport. Perhaps a gentle reminder about the value of the airport to the Island com-

A Trident in the livery of Channel Airways, which pioneered jets but collapsed in 1973

munity didn't go amiss, particularly when all too often Mr Mellor and previous airport administrators had had to justify more public spending, more claims on the surrounding countryside and more pieces of technical wizardry being situated sometimes in the unlikeliest of places.

There had been, for example, the new carport and freight handling area, built next to the main Les Quennevais road in the late 1960s at a cost of £80,000. There had been the up-to-date instrument landings system, installed for nearly £200,000 in 1969. And constantly, during the 1950s, 1960s and 1970s there had been the problem of land.

At one time, soon after the airport was built, a road ran across the eastern end of the airport perimeter at such an angle that when an aircraft was about to descend, a traffic light system halted all cars in the same way that a railway crossing halted vehicles to allow trains to cross (these were the first traffic lights in the Island and were operated from Air Traffic Control).

The road had to go (which it did) and, as time went by, more roads (for example, Rue des Landes), also went in the interests of air (and road) safety. But, as one road went, another would be needed to take its place and the network of roads currently hugging the airport's boundaries contains more than a few which were built to allow access to the heart of the Island.

Roads can always be moved, widened or closed (unlike the airport) which is one of the reasons why a new road, 1,200 feet long and running from the Pont du Val, via the Mermaid, to the roundabout at the eastern side of the airport was built in 1965, four years before James Sutherland's 12 perches of land were bought by the States after a compulsory purchase order had been placed on them.

Describing the State as 'judge, jury and jailer' at the time, because he had to agree to the sale, no matter how vehemently he protested, he was one of many parishioners whose grumbles began in the 1930s, in the mid 1960s, died down in the 1980s and came crashing back to a crescendo in the 1990s when further development plans were discussed.

Included in those grumbles from the 1970s were complaints about La Forge, St Peter, bought by the States from a reluctant Mr de Gruchy who didn't want to sell.

Complaints, too, about the parish football pitches, now owned by the Harbours and Airport Committee (on behalf of the States) and leased back to St Peter.

And the oddest complaint of all, voiced

The arrivals hall as it was back in the early 1980s

in 1979, about a new, £160,000 weather radar system installed at the northern end of the airport's boundaries. The complaint (and the reply) emerge when Senator Bill Morvan's soothing words are given:

'There is no fear of anyone being cooked by airport radar microwaves.'

(In 1979 no-one seemed quite sure what microwaves could or couldn't do).

By the 1980s and with a new president in charge of the Harbours and Airport Committee (Senator Bernard Binnington) and with a new commandant at the airport (Michael Lanyon) grumbles about new equipment, larger, noisier aircraft and unnecessary land-grabbing died down.

But as we will see, the Island — and the airport — doesn't stand still, and in 1994, when discussions were held to improve the airport yet again, the parish made its views on the matter known to anyone who cared to listen.

They would have welcomed a separate proposal, a few years earlier, that the entire airport be moved to the east — to the coast of St Clement — an idea which was voiced in the States and then forgotten about, because of the cost.

Parishioners from St Clement wrote to the 'Jersey Evening Post' at the time, alarmed that they would have to endure the kind of noise and large-scale building work St Peter had endured, but other parishes, too, had to give up their land for the sake of incoming and outgoing passengers.

In November, 1981, for example, a 100 foot wide navigational weather beacon was installed in a field at St Martin, much to the annoyance of neighbours, who complained that it was an eyesore. But it was necessary and the Island Development Committee agreed that it would be built on condition that it was adequately screened from public view.

Another complaint was effectively stifled in November, 1982, when the IDC agreed that the three existing aircraft noise zones straddling the airport and St Peter should be made longer and narrower to conform to a new noise survey. This was good news to the parish, because it meant that more homes could be built, and showed how consultation between parishioners and authority could benefit St Peter and the Island.

Jersey owed then and continues to owe a great deal to its airport and it is no coincidence that as it has grown, so Jersey has gained in prosperity and so, too, have the surrounding parishes of St Brelade and, of course, St Peter.

A Risky Business

'IT is fairly certain that were there no air services, hardly one quarter of the present number of tourists would make the journey by sea. Thus it is that Jersey's prosperity and even her standard of living become ever more dependent upon the maintenance of efficient airline services to and from all parts of Britain and upon the growth of those services as public demand warrants.'

('Evening Post' leading article, February 20, 1957)

How times had changed! At one time the bête noir of the airport and all it represented, the Island's only daily newspaper now recognised its contribution to the economy and to the tourist industry.

Without the airport, the increasing number of airlines wouldn't thrive and, without the airlines, new routes and new holiday-makers couldn't be found. The airport was doing well — but what about the airlines?

Until August, 1979, BEA (reconstituted into British Airways) jealously guarded their rights to the air routes over the Channel Islands, although other companies persevered, chartered their planes and discovered new destinations and passengers in both Britain and Europe. Airlines COULD survive if they worked on a shoestring budget and cut their cloth accordingly.

Maldwyn Thomas discovered that in 1948, when he founded Jersey Airlines and operated chartered flights between Jersey and France.

By 1949 he had scraped enough money together to buy his first Rapide (G-AKNF) and, in between farming and then being called to the airport to work at the counter, he took reservations, carried baggage and attempted to free himself of some of the restraints imposed by a Labour Government which expressly forbade private enterprise to enter any other sphere of civil aviation except charters.

By 1951, and with a Conservative government in power, BEA's control of the skies was relaxed. Air companies were permitted to join forces with BEA as associate members (this meant that BEA would be shareholders in your aviation company) and, by return, you could try to outbid them for routes to and from the UK.

Thomas did so, applying for licences to fly between all of the destinations BEA operated from Jersey and, in April, 1952, he pioneered the Jersey to Exeter route as well as establishing a regular Jersey to Paris service, first at Le Bourget and then at Orly.

By the early 1960s, Jersey Airlines were running their own fleet of Dakotas and brand-new Handley-Page Dart Heralds not only to Paris and Exeter, but to Amsterdam and to the Canaries (via Gatwick) as well. Started on a modest income it was now a million pound business machine and its rapid rise to success must have reminded more than one islander of a forerunner — Jersey Airways, killed by Socialism more than a decade before.

In 1960 Jersey Airlines was given the opportunity to do even more business, when the Conservative Government's new Civil Aviation Act received its Royal Assent. This meant that BEA's monopoly of home and European routes would end, and Jersey Airlines decided to buy back the 25 per cent stake BEA had in their company, and take their planes and their business to the northern hangar, where they stayed for most of the 1960s.

If the new legislation allowed Jersey Airlines more freedom, other air companies and potential rivals of both JAL and BEA were also off the lease. They included British Midland (formerly Derby Aviation) who were running regular Jersey to Castle Donnington flights by 1966.

They prospered, as Jersey Airlines had done, because they kept to their 'home' base and maintained a local image — a lesson more than one company operating in Jersey had failed to learn. In 1962, seeking more routes and keen to expand, Jersey Airlines became part of British United Airways. Maldwyn Thomas became managing director of Airlines (Jersey) Ltd, now operating under the auspices of BUA and other Ex-JAL people took top jobs, including Captain B W Gardiner, who became director of operations for BU (Manx) Ltd, and chief engineer Chandler, who became director of engineering for both companies.

The holding company called, suitably enough, Air Holdings Ltd, had many other aviation and holiday interests, including Morton Airways, Silver City and Transair, but if Mr Thomas thought that taking his mainly seasonal company into the shadow of a larger one would mean more, improved business, he was sadly mistaken.

A system of multi-user licences agreed with the Air Transport Licensing board allowed airlines within the larger group to operate on any other group member's licence so, from 1962, BUA began to switch their planes from route to route. This meant that, although nominally still a Jersey-based company, BUA (Jersey) was beginning to have its roots disturbed. Bigger might not necessarily mean better, despite the prospect of jet aircraft arriving in the Island for the first time.

By 1965 Maldwyn Thomas had resigned from BUA. He had other business interests in Jersey to look after, including a travel agency and links with Dinard-based Rousseau Aviation, who were, for a short time, an influential French/Jersey company in the early 1970s. Thomas also had political ambitions. By 1966, as Deputy of St

A BUA BAC 1-11. The airline grew out of Jersey Airlines but ran into difficulties

A Trident operated by BEA, which eventually became British Airways

Peter, he was speaking vehemently against noisy jet aircraft and longer runways. A year before this, he had been part of the company which had brought the first jet passenger plane to Jersey (in October, 1965).

The irony wasn't wasted on BUA chairman, Sir Myles Wyatt: 'Maldwyn Thomas had been active in promoting the jet service right up till the time of his resignation.'
(January 28, 1966)

He might have done, but a rebel turned is a dangerous adversary, and there is little doubt that he cared enormously for his adopted parish. The number of skirmishes Deputy Thomas and Senator Krichefski had, both inside and outside the House, is legendary, and although the Senator won most of them, the Deputy kept plugging away, eventually gaining rich reward. Without his campaigning it is doubtful whether the rules would have been relaxed to allow the St Peter development of new homes to go ahead, and his parishioners still thought highly of him in the early 1970s, when he was re-elected to the States, where a more relaxed, less abrasive style of airport management was being practised by the Harbours and Airport Committee and its president, Bill Morvan.

Sadly, by the 1970s, the company that Maldwyn Thomas had formed, had taken into (and out of) BEA and had then left, when part of BUA, was no more. BUA (CI) had ceased trading.

At the same time that Channel Airways brought their first 'big jet' into Jersey (a 139-seater arriving from Stansted in July, 1968) BUA were a worried company.

Maldwyn Thomas had once said that if you buy jets, you must look for longer routes and not plump for shorter, seasonal domestic destinations (like Jersey). They were only practicable all the year round and on longer runs, which suggests an expanding air industry, with many, regular routes available and fuel not forever going up in price, by leaps and bounds.

Unlike many other industries however, the world of aviation changes from year to year and so it proved with the fortunes of BUA (CI). In the autumn of 1968 the company announced that they would merge with BU(Manx). 250 people would lose their jobs.

'The current fleet was too large for off-season requirements,' the soon-to-be-redundant workers were told, and negotiations continued as management told the press that they hoped to continue the Jersey operations by running 'the right aircraft on the right routes with the right men'.

The phrase 'negotiations continued' can hide a multitude of sins and BUA redundancy plans were referred to the States, to BALPA (the pilots' union), to the CI Air Advisory Council, to Alan Bristow (one of the BUA management team from the UK) and to Rene Liron (local representative of the Transport and General Workers Union).

It was a terribly acrimonious dispute. Early in July, 1968, 30 BUA staff were given a ten per cent rise.

Three days later there was a strike by BUA engineers and general manager, Captain G Thomas, who had been vilified at a St Peter's Parish Hall meeting at the height of the jet controversy, was in the thick of the action again as he suggested: 'The best way of mitigating losses might be to cease operations.'

By the end of July, 1968, BUA (CI) did just that. Notice was given to another 500 employees at the firm and everyone from BUA passenger to BUA pilot was on tenterhooks . . . What would happen next? As other airline companies helped to ferry passengers booked on BUA flights back to the mainland, on August 14, at the height of the summer season, the dispute was settled. It cost BUA at least £150,000 and 23 days of anxiety before redundancy terms were agreed, and there was a restructuring of all BUA services before BUA (CI) was formally wound up on October 31, 1968.

The public were told that, at one stage several months before the end, the total assets of the firm had been £526, and although BUA (Airways) emerged from the mess, it took a long time before the bitter taste of the dispute was washed away from the airport forecourt.

A Change of Colours

AFTER every storm there is always a gentle breeze and, as the stormy scenes of BUA(CI)'s death throes lingered over the tarmac, a few bright, breezy reminders that airlines CAN make money out of Jersey Airport drifted into view. On February 18, 1969, Captain D H Stuart, A I Le Gresley, L C Thomas and G L Gillborn (all ex-BUA men) launched Intra Airlines with one DC3 (Dakota) and a lot of hope.

Six months before this, an advertisement had appeared in the 'Evening Post' offering daily services to Alderney, Granville and Guernsey. By operating a policy of cash-on-the-nail, no-credit facilities (it kept book-keeping to a minimum) and a 'walk-on, walk-off' service, Aurigny Air Services Ltd began in a small way but, by July, 1969, were operating seven Islander aircraft.

Despite the odd opening times of Alderney airport (the only civil airport in Europe to close between 12.30 and 2.00 pm for lunch), by the end of 1969, under the careful control of general manager Fred Morton, Aurigny had a staff of 54, eight Britten-Norman Islander aircraft and a total number of passengers for the year of 105,000. Although they relinquished their 'walk on, walk off' policy four years later, as security tightened and as the number of flights they operated increased, by keeping air fares as low as possible (in 1976, for example, as larger companies pressed vigorously for increased fares, they told the CAA that they were reasonable happy with their own inter-island prices) they prospered.

In 1972 they opened their own passenger terminal and when, in 1978, they were taken-over by Anglo-Normandy Aviation, it wasn't a 'bailing-out' operation — Aurigny were doing too well for that. By keeping their ambitions relatively small, by 1991 their annual turnover was £8 million.

A year later they had proved their worth well enough to win the contract to bring the Royal Mail into the Island from Gatwick (it had previously been Southend) and, two years later, it was announced that they were going to introduce a bigger plane, the 36-seater Short 360, onto their Channel Island routes, to add to their fleet of Trislanders. The company, now owned by Exxtor Ltd, continued to expand and in 1997 introduced a Jersey to Caen route to complement their other French routes and was carrying 300,000 passengers a year.

Meanwhile, what of BEA, who must have been intrigued by the ups and downs of BUA's fortunes, even when Maldwyn Thomas hinted that he would step in, take 'Jersey Airlines' out of the company and start again?

BEA had problems of their own — and what had happened at BUA was enough to send shivers down many an aircraft spine. They were committed to scheduled flights to the mainland, knowing that Jersey was a seasonal destination and that other companies were looking at their more lucrative routes with greedy eyes. Despite the many passengers they carried (which helped to make Jersey Airport the sixth busiest in Europe during the 1970s) there was little profit in their Channel Island route.

In November, 1970, as the White Paper on the Edwards Report on air travel was being debated, BEA announced that they 'expected a £700,000 loss next year, even with an increase in fares'.

The Civil Aviation Authority wouldn't allow them to increase prices as they had hoped, then on February 4, 1971, BEA confirmed that their services would include the setting-up of a company trading as BEA (Channel Islands) within the BEA/British Airways Services group.

It was also proposed that Viscounts, rather than true jets, would be used on the Channel Islands route, prompting criticism from Tourism president Clarrie Dupré who preferred the jet engine. How times had changed! But, with other air companies now competing hard for routes, and with all-inclusive charters of flight, accommodation and meals at reduced rates now permissible (which was found to affect scheduled flights), BEA were struggling.

As the receiver was brought in to sort out

Channel Airways (who had carried 85,000 people to Jersey in 1971), BEA were contemplating merging with BOAC, which they did in January, 1973. This, however, was a crisis period in the UK, for if Channel Airways and BUA(CI) could go bust BEFORE the three-day week, increased oil prices and the miners' strike of 1974, what would be the effects immediately AFTER this burst of economic catastrophe?

Britain's leading tour operators were predicting that it would be a gloomy summer; the CI Air Advisory Council were backing BIA's plans to end British Airways' (formerly BEA/BOAC) monopoly of the Jersey to London route; 40 maintenance staff at BA lost their jobs and, as BA announced a 15 per cent fare increase in September, 1975, marketing manager Alan Wright claimed: 'In 28 years of operations on the route we have made a profit only about four times.'

A single fare to London became £31.80. By comparison, in 1970, a midweek return cost £13 9s 8d (just under £13.50).

At first glance it appeared as if the travel trade was going through a minor crisis. In reality, however, although one or two companies were experiencing difficulties, the number of passengers actually using the airport in August, 1975, was 104,085, an increase of 650 on the previous year.

In 1978 'the face of rising air fares' had changed dramatically. After BMA contested CAA regulations and after the break-up of the Memorandum of Understanding between airlines, which normally resulted in a gentleman's agreement about fare-fixing, a mini-price war raged.

Ipex, Apex and various other cut-price fares tempted people to come to Jersey, and at one time a return trip to London was 40 per cent cheaper than it had been a year before. Freddie Laker's influence, to make air travel as cheap as possible, pervaded the skies, but air figures didn't rocket during 1978, and although that year they made the front page of the 'Evening Post' when a 164-ton BA Lockhead Tristar, the biggest commercial plane ever to arrive in Jersey, cleared in one gigantic go a backlog of passengers, a year later they announced dramatic changes. They were going to cut their routes to the Channel Islands.

Although the management made uncompromising statements like 'we will not dissipate scarce resources on unprofitable routes' it wasn't as bad as it sounded, especially as it opened the door for other companies to try their luck on the Channel Island route.

The Channel Island Air Advisory Council recommended that British Midland, Dan Air, Jersey European Airways and Air UK should split the Jersey to Southampton, Birmingham/East Midlands, Cardiff/Bristol, Stansted, Glasgow, Edinburgh and north-eastern routes between them as British Airways kept the rest, scrapped all cut-price fares and, with a sigh of relief, for the moment slipped the reins.

That 'for the moment' is important in the story of British Airways' involvement in airport history, for by the late 1980s they were back with a vengeance, and it was a sign of how times had changed when, on October 13, 1994, BA Channel Islands manager Peter Crespel announced that for the first time in 50 years BA would begin a direct Jersey to Paris schedule. Similarly, when Crespel presented a certificate to Bob Wickings, managing director of Executive and Business Travel in June 1994, marking his company's sale of £2 million worth of BA tickets, there could be no doubt that by the 1990s British Airways had made a substantial investment in the Island, and as well as servicing both Gatwick and Heathrow could then offer passengers direct flights to several other UK cities, and to Dusseldorf, via London.

Meanwhile, what of Intra, whose first scheduled flight had been between the Channel Islands and Gloucester and Cheltenham, in May, 1971?

In 1972 they announced profits of £180,000 and, by the end of the year, added a third DC3 Dakota to their Jersey-based fleet. The previous year they had carried 20,000 passengers and 125,000 kilos of freight; by 1978 Intra owned the largest fleet of Viscounts available on charter in Europe. They had a staff of 140 and a turnover of £4 million. Things were going swimmingly (or

British Caledonian took over BUA and flew to Gatwick

so it would appear) until one day in November, 1978, when the police invaded Intra's airport offices and issued the following statement: 'Police are investigating suspected breaches of a number of statutory laws relating to Intra Airways'. One of those 'breaches of the law' involved Intra's pilots flying more hours then permitted, but despite mounting problems, Intra Airways had potential. Express Air Freight (CI) realised this in late December, 1978, when they combined with Intra in what was described as a 'bailing-out operation'.

In 1979, the company bounced back in a new guise, as three businessmen from three very different walks of life took over the company, rechristened it Jersey European Airways and helped to restore its fortunes. The company was helped in its ambitions in 1988 with the collapse of Air Europe, which flew Guernsey to Gatwick.

JEA took over the route, and introduced a similar Jersey to Gatwick run. The owner of the company by this time was Jersey resident Jack Walker, who said: 'We are intent on making the Channel Islands our central focus. Other airlines want to put their profits back into developing overseas routes and UK hubs — we want to reinvest in Jersey . . .' Although they had to concede the Jersey to Dinard route to Aurigny in 1994, by 1997 they were looking to expand, and were offering Islanders a direct Jersey to Dublin service.

British Midland flew their millionth passenger in 1970, and after operating Jersey's first, all-inclusive package holiday from the UK, were later granted an application to extend winter-only services from Birmingham into the summer.

British Airways objected but, six years later in 1977, it was British Midland's turn to object as Dan Air claimed the right to the potentially lucrative Channel Islands to Newcastle route. Then, as British Airways revealed that they would be pulling out of some of the less economical Channel Island routes in 1979, British Midland were one of the first airlines to step in to claim them as their own, and they became one of Jersey Airport's main customers, although their 'home base' has remained Castle Donnington (East Midlands).

And BU(I)A? They cut back, operating only one BAC 1-11 to Gatwick, daily, while looking for new direction. In 1969 they began a service to Swansea and, in the same year, they pulled out of the Jersey to Alderney route on the basis that their De Havilland DH 114 Herons were uneconomical on such a short flight.

In July, 1970, they left the BUA group and a month later their new livery was prominent at the airport. They brought in the papers and mail, ran a one-way postal service to Orly airport, Paris, and in 1977 their management revealed that they were not prepared to buy jets any more because they were of no practical use on the routes they operated. They attacked BA for monopolising the best airways over the Channel (in December, 1976, 347 BIA staff petitioned the Secretary of State at the Department of Trade on the grounds that the CAA gave BA preferential treatment) and, in 1979, joined with Air Anglia and BIA/Air West in preparation for the route-release BA had recently promised.

The company, renamed Air UK, took their chances, and when Guernsey Airlines' parent company, British Air Ferries, went 'into administration' in January, 1988, thus relinquishing their Jersey to Southampton route, seized the initiative, and by the 1990s were flying to the south coast as well as to other airports, further north, and, via the UK, to Europe.

Other airlines who made a commitment to the Island included Crossair, which in the 1990s flew every day direct to Zurich and, in association with BA, Manx Airlines. However along the way there have been casualties, including, in the decade leading to the 60th anniversary of Jersey Airport being built, Air Atlantique.

Meanwhile, one successful 'airline' that ought to be mentioned is a club — the Aero Club — which flourished when, in November, 1951, a handful of islanders keen on flying met at the Old England Hotel, Cheapside. Without premises or a plane they formed the Channel Islands Aero Club, elected a chairman (Captain Tommy Froggatt, a BEA pilot) and rented a club-house (a room on Patriotic Street).

It was hardly the best place from which to run an aero club, but within three years they had acquired a small, timber red-and-white hut near the airport's playing fields.

By 1955 they had acquired their own, single-storey clubhouse which looked out on to the flying field where, parked outside on the grass, was the club's own Tiger Moth. A new Auster had been ordered and to complete the early success, Wing Commander M. Pickford DFC, was taken on as full- time flying instructor. He served the club for the next eight years.

The Aero Club continued to prosper. Jersey Airlines loaned them a hangar, helped to maintain members' planes and, after Pickford retired, provided an instructor. By 1964 40 local pilots had obtained their licences. There were 430 members and, for the last ten years, the CI Aero Club had hosted one of the most prestigious air events in Europe — their international air rally. And the club had not stopped growing. They built their own clubhouse at the airport, and by the 1980s owned three Cessna 152s, two PA-28 Archers, one PA-28 Warrior and a Chipmunk.

BIA, eventually to become part of Air UK, operated BAC 1-11s

The Eighties

THE 1980s came quietly to Jersey Airport, and despite a few murmurs through the years, stayed that way.

SPAD survived, but fewer meetings were called at St Peter's Parish Hall to discuss aircraft disturbances.

For in the 1980s the problem of noisy jet aircraft was tempered with new developments in aircraft technology. Jet aircraft didn't need to be noisy, which is one of the reasons why, in January, 1980, when British Airways announced that they intended to operate all-jet services in 1981 and to replace the BAC 1-11 on the Jersey route by the Boeing 737, Deputy Michael Bonn, a member of SPAD at the time, said: 'This is the best news I have heard for a long time'.

As we will see, however, it wasn't always peaceful between parish and airport in the 1980s or the 1990s when it came to land development. In March, 1981, for example, the States were told that they would have to buy Archlow Villa, in St Peter, by compulsory purchase order. In the States Members were told that it was too close to the runway, and an aircraft was once said to have knocked the aerial off the roof of the property as it flew overhead. It was bought for £50,000.

However, one of the main battles over the airport in this decade was not at ground level, it was in the sky above.

In November, 1979, Senator Dick Shenton launched a stinging attack on the Channel Islands Advisory Council, whose chairman of the Jersey delegation was Senator Bill Morvan.

Shenton spoke about the council's lack of teeth as it was announced that BA were pulling out from scheduled routes apart from the Jersey to Heathrow and Jersey to Manchester runs, adding that the CIAAC made excuses for BA, instead of looking after the Island's interests.

His fury and frustration were understandable, although he knew, as did all States Members, that the CIAAC was as toothless then as it had been when it was formed in 1947, despite recent amendments to the regulations which allowed the council to ask for hearings to be suspended if anything new was brought up, to give them the chance to comment.

'Comment' is very different from 'instruct' or 'command', and throughout the 1980s Britain's Civil Aviation Authority had the power to ignore the Channel Islands completely and to allocate air routes to any company it chose.

The question of routes was to surface yet again, in August, 1982, when a joint delegation of States Members from Jersey and Guernsey, headed by the Bailiffs of the two islands, Sir Frank Ereaut and Charles Frossard, flew to London to meet key Government figures, including Lord Elton, the Minister of State at the Home Office with special responsibility for Channel Island affairs.

Their aims were to achieve some sort of better deal, to have more say in decision-making, as Senator Shenton, one of the Jersey delegates, said at the time, and by 1986 constant pressure on the British authorities seemed to have paid off. For on February 1 that year it was announced that a working party consisting of officials from the Home Office, the Department of Transport, and the Jersey, Guernsey and Alderney, had recommended that the islands should have a special position at the Civil Aviation Authority when they considered new air services for island routes.

At the time, the chairman of the Jersey delegation to the CIAAC, Senator Bernard Binnington, welcomed the proposals, which recognised the islands' special problems and unique position,.

Senator Binnington was cautiously optimistic, but in November that same year the Transport Secretary, Mr John Moore, reversed a decision made by the CAA, with the blessing of the CIAAC, to allow Dan Air sole rights on the Bournemouth to Channel Islands route. He gave JEA permission to compete on the route after the company had appealed against the decision, which suggested that the British Government

1987: The outline of the old building can be seen but the airport has doubled in size

reserved its right to interfere, despite the fact that the Island controlled a huge piece of the sky and all the planes which passed through it as well as all of the aircraft which land in the Island.

For the Channel Islands Control Zone includes all of the other Channel Islands from 49 degrees north to 50 degrees north (the Minquiers to a third of the way across the Channel and out to some 30 miles west of Jersey from near the French coast).

Initially the ceiling of the zone was 3,500 feet. In the 1960s it increased to 5,000 feet, then to 11,000 feet, and then, to accommodate jets, to 20,000 feet.

Any plane passing through this air space is air traffic control's responsibility, and 130,000 aircraft pass through it each year, earning the Island a great deal of money from the European Organisation for the Co-ordination of Safety (Euro-Control) for 'international aeronautical services rendered'. It is hard-earned money when you realise that ATC officers have to undergo rigorous, mainland training before returning (the pass rate is around 70 per cent).

As well as positioning aircraft to land (about 50 an hour, on average), and sending them through different airways, to ensure that each is at a correct height and course, their duties extend to funnelling overhead planes into airborne layers called airways. These are described by colours and numbers, and one of the busiest is Airway Red One (used between Jersey and London) which starts in Spain, finishes in Russia, and passes over 101 countries.

A separate but lesser fuss in the 1980s was over the installation of instrument visual range equipment in July, 1982, at a cost of £180,000. For initially the equipment was not used, and the old way of measuring visibility (firemen would count the number of lights they saw on the runway) continued.

But this kind of criticism, against the installation of equipment, including a new £1 million instrument landing system in 1986 and a recent upgrading of Air Traffic Control's displays and communications system, costing over £1million, was nothing to the fuss that was about to break out over plans to build a bigger, better airport. That argument, about its need, cost and positioning, was to dominate the 1990s.

Heading to the Millennium

ON 11 January, 1997, Deputy Jimmy Johns, president of the Harbours and Airport Committee, was quoted in the Jersey Evening Post as saying: 'If you asked the people of the Orkneys, Shetlands, or any other island community what they thought of our transport links and the range of fares, then I can tell you they would be envious of Jersey.'

He was making a valid point at the time, for how many other islands of similar size to Jersey can anticipate between 1.7 and 1.9 million people arriving each year by air?

Those were the figures that his predecessor, the late president of Harbours and Airport, Deputy John Le Fondré, had quoted in 1994, when he was explaining the need for between £15 and £17 million to be spent on redeveloping the airport site.

Fifty-seven years before, he said, no-one could have anticipated the demands of what was then a comfortable, provincial airport. Nor could they have envisaged an age when every passenger coming into the Island, or departing from it, would have to go through a stringent security check following the Lockerbie disaster on December 21, 1988, when a terrorist bomb exploded in a Pan Am Jumbo jet over Lockerbie, killing all 259 passengers. On the ground another 11 people were killed by the falling débris.

'New security measures which need to be installed mean that the building is creaking at the joints,' Deputy John Le Fondré said, before going on to describe what happened when the lounges became so full to overflowing, that passengers had to wait in a large airport marquee when planes were delayed by fog.

'I spoke to one old lady who was in the tent and who said she was very uncomfortable and embarrassed for Jersey,' he said, adding that over the years the airport had earned money for the States, and it was now time that the States repaid that favour with interest.

It was February when he talked about his intention to ask the States to help pay towards an up-to-date, larger airport. And although the press sniped at the costs, which were to increase steadily over the next three years from a 'possible' £17 million to a 'probable' £22 million – perhaps more – everyone in the Island, particularly those who had been delayed by fog and had curled up to sleep on a hangar floor or at a nearby sports hall, knew the airport was large enough to cope midweek, during the winter, but not so good when 20,000 passengers might pass through the airport doors one Saturday in summer.

Headlines in the 'Jersey Evening Post' like: 'Air passengers to wait in hangar' from July, 1988, or 'Airport Fog Nightmare' five years later and 'Scandalous to be herded like cattle' in June, 1994, were a constant reminder to airport chief Michael Lanyon that this was not the kind of advertisement that Jersey, which relied so much on tourism, could afford. Nor was there any ready-made solution, other than spending a lot of money to improve the facilities.

Lanyon had said as much in March, 1994, when talking about how a £35 million plan to completely rebuild the airport had been rejected by the purse-holders of the States, the Policy and Resources Committee, in 1989, on the basis that it was too much — but that a reasonably-priced scheme might meet their approval.

The idea of a cheaper scheme had been firmly planted in Deputy Le Fondré's mind by his predecessor, Senator Bernard Binnington, who had proposed a £12 million modernisation package in 1993. But Deputy Le Fondré, as Bernard Binnington before and Deputy Jimmy Johns who succeeded him, knew how loath the Island might be, to spend so much money on a mere gateway to Jersey — but might there not be a cheaper way, to find the money?

Lanyon and Deputy Le Fondré, when the revised scheme was placed before the House, believed the answer was 'yes'.

The States would fund a proportion of the costs, Deputy Le Fondré said, but the

BA brought Concorde to Jersey to mark the airport's 50th birthday in March, 1987

rest would be found in the commercial sector. Lanyon, when interviewed by the 'Jersey Evening Post', explained further.

Over the last 16 years, he said, he had received offers of loans totalling £100 million from the private sector from about 17 companies interested in investing in the airport, as long as the States contributed an equal share. He added that over that time the airport had made a £20 million profit, and that a new, larger departures terminal would lead to more franchises, including a new bar, shops and fast-food outlets. 'If the States contributed £10 million,' he said, 'the airport authorities would find the rest'.

The States were effectively being asked to become partners in a States/public owned company, a kind of 'Airport PLC'.

Initially, however, States members were not impressed; particularly after being asked for a £10 million grant 'not a loan' when a 'no-strings attached gesture to the redevelopment of the airport' was asked of Policy and Resources early in 1994.

The committee said it was not a priority — which effectively meant it would be the next century, if at all, before the redevelopment went ahead — but Deputy Le Fondré wasn't to be dissuaded, and in March that year he persuaded the committee to agree a £10 million loan (not a grant) to the airport, provided they find the rest of the money.

'We are delighted to have this money offered to us, even though it has still to go to the States,' Deputy Le Fondré said.

He was obviously worried about how the money would be found to complete the buildings and repay the loan, a concern echoed by Peter Crespel, manager of British Airways in Jersey, who said: 'we hope the funding can be managed without the necessity to pass on the costs of development to the passenger through higher landing and passenger charges.'

He was right to be worried. For on July 12, 1994, Deputy Gary Matthews asked the States to approve a departures tax on all flights out of the Island of between £1 and £2 per person. His proposal was received with dismay in the offices at Tourism and with incredulity in the offices of the airlines who flew in and out of Jersey. For while Deputy Matthews was only following the precedent set by UK Chancellor Kenneth Clarke, when he had set a £5 departure tax on all those people leaving the UK by air (but not, paradoxically by sea) he was effectively dissuading potential holidaymakers who might have thought of flying to and from Jersey.

As the president of Tourism, Dick Shenton, said at the time: 'I cannot imagine that the public reaction would be favourable . . . the biggest problem with Jersey is already the high air fares . . .' Deputy Matthews' proposition came to nothing, but not before a healthy debate took place in the States about the airport's future. In effect it was a minor distraction from an inevitable conclusion for in May, earlier that year, the States had agreed to finance a new airport's development plan. Less than a year later, on 25 March 1995, the Jersey Evening Post reported as follows: 'The two-day exhibition of the proposed £20 million airport project ended yesterday having attracted more than 1,000 viewers — most of whom supported the ideas.'

Those who were less than keen to have a larger airport included many of the people who lived in the parish, including Mac Pollard, the Constable of St Peter, and St Peter's Deputy, Tom du Feu, plus the conservation group, Save Jersey's Heritage, who were angry not so much by the size of the improved airport, more by the way it might look. Not content with just voicing an opinion, they approached UK architects Terry Farrell and Company who drew up alternative plans which, they claimed, would cost no more than £10 million.

Marcus Binney, chairman of the group explained why the States should think again, by describing the 1937 main building as: 'rare and historic', needing 'something extremely elegant' to be built alongside.

He said that according to the current plans, 'the new departure hall beside it' had the look of a 'glorified shed.'

When the deputy-chairman of the Harbours and Airport Committee, Len Norman, protested that new plans would

By the 1990s Jersey European were offering direct flights between the Island and Ireland

A British Midland DC 9 arrives from East Midlands Airport, on the outskirts of Derby

'appear to be more expensive than the plans that we are going ahead with' Binney said he did not believe this was the case and repeated his message, that the current plan the airport were working towards which featured a canopy surround and extensions to the 1937 building would 'disfigure' it. It was to be another month before the group conceded defeat, and only then because 'slender finances' meant that they could not afford to press ahead with their own plans.

Even so, they went down fighting, with Binney arguing that the Harbours and Airport Committee had 'missed the opportunity' to preserve the look of the old building. Nevertheless, the airport's own design, drawn up by a Public Services team led by chief architect Jon Kempster, was eventually accepted. The final airport package included a grant of £11 million from the States and private finance in the form of a loan of £12 million from Midland Bank International Finance Corporation/HSBC Investment Bank plc wasn't approved until June, 1996, by which time building work had begun — and was then stopped for a while, when structural, mechanical and electrical engineers Russell Wilson International, who had been involved with the airport development since 1994 went en désastre (bankrupt) in March, 1996, but after a pause in building work and after a change to the original plans redevelopment continued as before.

There were to be other hiccoughs along the way — British Midland passengers, for example, found that they had to check in nearly half a mile away from the main departures hall, and would be ferried to the airport by minibus, but anyone passing the site could not fail to notice that a major overhaul was going ahead, by night as well as by day. In years to come it is inevitable that the 'new' airport will be linked with the one constant factor during its conception, through to the opening of the new doors, airport director Michael Lanyon, who had pressed for improvements a decade before the main part of the refurbished airport was opened in March, 1997.

To the airport authorities, a larger airport meant happier customers, and customers would be unhappy if they had to queue to go through Customs or airport control, or had to sleep on a chair in a tent when it was foggy, because the departures hall was far too small. However, if fog at the airport always made the news — and it is reckoned that it cost the Island £80,000 a year to deal with flights and passengers when the fog closed in — the airport was often newsworthy in his own right.

On Thursday, July 8, 1993, for example, skips and barriers were placed outside Aurigny's offices to blockade the aircraft manoeuvring space, because the airport authorities were not convinced that Aurigny Air Services were authorised to act as handling agents for Newquay Air which had flown to Jersey on an inaugural flight that morning. Lanyon was concerned about airport security at the time, for Aurigny had, both then and for many years beforehand, run a separate terminal, one which didn't enjoy the same strict security controls of the airport 'proper'. The authorities were worried that passengers and luggage arriving from the West Country could slip into the Island unnoticed and without challenge.

Lanyon later apologised for ordering the blockade, but the point had been made — although Aurigny were the Channel Islands own airline, they had to fall into line, so that there should be no loophole which could allow terrorists, or drug-pushers,

Travellers come first: A passenger service vehicle

easy access to Jersey. The Lockerbie disaster had made everyone security conscious. Special dispensation was not given to Aurigny to handle luggage brought in by Newquay Air.

This was 1993, five years earlier he had made the news when in December, 1988, he warned of a possible confrontation between management and air traffic controllers when the latter said they would no longer supervise trainees, because they felt they were underpaid.

The dispute was eventually settled; and in retrospect it probably wasn't as threatening as when, on August 27, 1994, the airport fire crews conducted a 24-hour strike (at a time when fog had caused chaos enough at the airport), again over pay. Once again Lanyon was called upon to make a statement — although it was Establishment, under the presidency of Deputy Frank Walker which was effectively calling the paymaster's tune. The dispute was settled but both sides were back in the news in August, 1996, when, yet again, the fire crews threatened action before the matter was effectively resolved.

An Aurigny Trislander comes in to land

replace them. Transport and General Workers Union leader Mick Kavanagh was called in and the matter was finally sorted out, but not before the dispute had made the headlines on Channel Television and in the 'Jersey Evening Post'.

The two men were used to having a high profile in the community; but then so, too, had Kavanagh's predecessor as local official of the T&GWU, René Liron. In February, 1986, for example, he had hit the news when it was announced that British Airways were building an executive lounge for its up-market passengers.

'Why can't they rub shoulders with me?' he said. 'It's a waste of area and it's not necessary.'

Disputes — over pay, over conditions, over passengers — will always be a possibility, but while the staff at Jersey Airport have had their own problems to contend with, so too have the airlines, especially when having to bid against other airlines for air routes in and out of Jersey.

These routes, as we saw in the last chapter, have traditionally been controlled not by the Channel Islands, but by England, whose Civil Aviation Authority issued the appropriate licences.

Licensing has been a vexing subject, and one which prompted Craig Alexander, managing director of Aurigny at the time, to criticise the CAA during the airline's 25th

Indeed 1996 was a year for unrest, for that same year it had taken nine hours of talks to finalise a deal over refuelling after Simon Engineering took over the franchise from Air UK, whose employees said they had not been properly paid.

That same year Jersey Airport nearly closed down, because the States-paid cleaners wanted an all-out strike when they heard that contract cleaners might

1997: Air UK maintain close links between Jersey and the south of England

February 16, 1997: The 'new' airport skyline, as contractors aim to be finished by early March

birthday party on Monday, March 1, 1993, for he had learnt that other airlines as well as Aurigny were to be granted inter-insular licences, even though the Jersey Transport Authority and the Guernsey Transport Board urged the CAA to overturn their decision because they were afraid of what they called 'predatory' carriers, interested in the islands' money, but not necessarily their welfare.

Aurigny won; just as they had won another significant battle in 1992, after they had put in a bid to fly the Royal Mail into the Islands.

Craig Alexander explained as follows: 'Freight and charter services have always complemented the scheduled passenger routes and mercy ambulance flights now average almost one a week . . . However we were delighted to be awarded the Royal Mail Charter to carry 80,000 letters each weekday between Jersey, Gatwick and East Midlands Airport and hope to expand this area of our business with the 360.'

The Short '360' Mr Alexander was talking about was the newest aircraft Aurigny had on the route — so large that it needed a stewardess, as well as a pilot, unlike 'Joey' their friendly 16-seater Trislander.

Good news, bad news

TOWARDS the end of January, 1997, it was announced that the Civil Aviation Authority wanted seven homes and the St Peter Football pitch to be removed to comply with the recommendations of the Civil Aviation Authority recently published on Jersey. The parish were up in arms.

As former St Peter Deputy Tommy du Feu explained: 'We have considered digging our heels in and refusing, but I am sure that they will just roll out the big guns and make certain we comply. The airport just seems to be taking up more and more of the parish of St Peter and using more and more resources of the Island of Jersey. We have only 45 square miles and it's time we started taking that into proper account.'

The report, on 24 January that month by the CAA actually gave the Island a complimentary report on the state of the airport . . . but it wanted changes. And the compilers of that report knew, full well, that the British government had already been very good to Jersey, in giving it something no other comparable piece of land the size of Jersey actually had — control of its own air space.

Throughout this history of flight in

British Airways have been associated with flights to and from the Island for over 60 years

Jersey the same complaint by Jersey and Guernsey authorities has been heard, decade upon decade — the islands should control who flies into them. Meanwhile decade upon decade the Civil Aviation Authority had reaffirmed its commitment to administer all air routes across Jersey.

However, when the CAA proposed deregulated skies in 1993, which meant that any airline could fly to any airport, the Jersey Transport Authority were agitated more so than at any other time in its history. That might be fine in summer, they said, but what about the winter?

Would an airline fly to Jersey in July and August, only to walk away from the routes after September 1st?

So in January, 1994, a draft report on route licensing was prepared and discussions took place with the UK Department of Transport for their approval.

As JTA chairman, Derek Maltwood, explained: 'Under the terms of the draft policy, the islands will issue a permit to any airline holding a route licence granted by the CAA unless they feel this service would be contrary to the best interests of the Island. A Jersey air licensing authority will allow us to protect our airline routes by making sure the airlines which fly here in the summer also have to operate winter services. We need sufficient flights both in the summer, for the tourists, and all year round for the islanders.' In other words, no airline could disembark passengers at Jersey Airport unless they were committed to the island's long-term welfare.

On July 4, 1994, the Department of Transport saw sense, and from 1995 it was agreed that Guernsey and Jersey had the right to licence carriers, according to Channel Islands needs. Not only was this agreement unique to Europe — it was also a smiles-all-round-victory to those people who had taken part in the negotiations.

With a larger airport planned for 1997, and cautious optimism in the travelling trade, this was a real success, not just for Jersey, but for all Channel Islanders.

Only one problem now remained — how much extra land would the Civil Aviation Authority demand before the airport (to meet safety requirements) stopped expanding?